Climate Change Indicators in the United States, 2012

TECHNICAL DOCUMENTATION

December 2012

Contents

Overview

This document provides technical supporting information for the 26 indicators that appear in the U.S. Environmental Protection Agency's (EPA's) report, *Climate Change Indicators in the United States, 2012*. EPA prepared this document to ensure that each indicator is fully transparent—so readers can learn where the data come from, how each indicator was calculated, and how accurately each indicator represents the intended environmental condition. EPA developed a standard documentation form, then worked with data providers and reviewed the relevant literature to address the elements on the form as completely as possible.

EPA's documentation addresses 13 elements for each indicator:

1. Indicator description
2. Revision history
3. Data sources
4. Data availability
5. Data collection (methods)
6. Indicator derivation (calculation steps)
7. Quality assurance and quality control (QA/QC)
8. Comparability over time and space
9. Sources of uncertainty (and quantitative estimates if available)
10. Sources of variability (and quantitative estimates if available)
11. Statistical/trend analysis (if any has been conducted)
12. Data limitations
13. References

In addition to indicator-specific documentation, this appendix to the report summarizes the criteria that EPA used to select indicators for inclusion in the original report, published in April 2010. This documentation also describes the process EPA followed to select and develop those indicators that have been added or substantially revised since the publication of EPA's first version of this report. All indicators included in the report met all of the selection criteria. Lastly, this document provides general information on changes that have occurred since the 2010 version of the *Climate Indicators in the United States* report.

EPA may update this technical documentation as new and/or additional information about these indicators and their underlying data becomes available. Please contact EPA at: climateindicators@epa.gov to provide any comments about this documentation.

EPA's Indicator Criteria

General Assessment Factors

When evaluating the quality, objectivity, and relevance of scientific and technical information, the considerations that EPA typically takes into account can be characterized by five general assessment factors:[1]

- **Soundness:** The extent to which the scientific and technical procedures, measures, methods, or models employed to generate the information are reasonable for, and consistent with, the intended application.

- **Applicability and utility:** The extent to which the information is relevant for the Agency's intended use.

- **Clarity and completeness:** The degree of clarity and completeness with which the data, assumptions, methods, quality assurance, sponsoring organizations, and analyses employed to generate the information are documented.

- **Uncertainty and variability:** The extent to which the variability and uncertainty (quantitative and qualitative) in the information or in the procedures, measures, methods, or models are evaluated and characterized.

- **Evaluation and review:** The extent of independent verification, validation, and peer review of the information or of the procedures, measures, methods, or models.

Criteria for Including Indicators in This Report

EPA used a set of 10 criteria to carefully select indicators for inclusion in *Climate Change Indicators in the United States, 2012*. The following table introduces these criteria and describes how they relate to the assessment factors listed above and the 13 elements in EPA's indicator documentation form.

Criterion	Description	Documentation Elements
Trends over time	Long-term data are available to show trends over time. These data are comparable across time and space. Indicator trends have appropriate resolution for the data type.	4. Data availability 5. Data collection 6. Indicator derivation

---

[1] For more information about these assessment factors and their application, see: U.S. EPA. 2003. Science Policy Council assessment factors: A summary of general assessment factors for evaluating the quality of scientific and technical information. EPA 100/B-03/001.

Criterion	Description	Documentation Elements
Actual observations	The data consist of actual measurements (observations) or derivations thereof. These measurements are representative of the target population.	5. Data collection 6. Indicator derivation 8. Comparability over time and space 11. Statistical/ trend analysis
Broad geographic coverage	Indicator data are national in scale or have national significance. The spatial scale is adequately supported with data that are representative of the region/area.	4. Data availability 5. Data collection 6. Indicator derivation 8. Comparability over time and space
Peer-reviewed data (peer-review status of indicator and quality of underlying source data)	Indicator and underlying data are sound. The data are credible, reliable, and have been published and peer-reviewed.	3. Data sources 4. Data availability 5. Data collection 6. Indicator derivation 7. QA/QC 11. Statistical/ trend analysis
Uncertainty	Information on sources of uncertainty is available. Variability and limitations of the indicator are understood and have been evaluated.	5. Data collection 6. Indicator derivation 7. QA/QC 9. Sources of uncertainty 10. Sources of variability 11. Statistical/ trend analysis 12. Data limitations
Usefulness	Informs issues of national importance and addresses issues important to human or natural systems. Complements existing indicators.	6. Indicator derivation
Connection to climate change	Climate signal is evident among stressors. The relationship to climate change is easily explained.	6. Indicator derivation 10. Sources of variability
Transparent, reproducible, and objective	The data and analysis are scientifically objective and methods are transparent. Biases, if known, are documented, minimal, or judged to be reasonable.	4. Data availability 5. Data collection 6. Indicator derivation 7. QA/QC 9. Sources of uncertainty 10. Sources of variability 12. Data limitations
Understandable to the public	The data provide a straightforward depiction of observations and are understandable to the average reader.	6. Indicator derivation 12. Data limitations

Criterion	Description	Documentation Elements
Feasible to construct	The indicator can be constructed or reproduced within the timeframe for developing the report. Data sources allow routine updates of the indicator for future reports.	3. Data sources 4. Data availability 5. Data collection 6. Indicator derivation

Process for Evaluating Indicators for the 2012 Report

EPA published the first edition of *Climate Change Indicators in the United States* in April 2010, featuring 24 indicators. In 2011, EPA began to develop a second edition using the following approach to identify and develop a robust set of new and revised indicators for the report:

A. Identify and develop a list of candidate indicators.
B. Conduct initial research; screen against a subset of indicator criteria.
C. Conduct detailed research; screen against the full set of indicator criteria.
D. Select indicators for development.
E. Develop draft indicators.
F. Facilitate expert review of draft indicators.
G. Periodically re-evaluate indicators.

In selecting and developing the climate change indicators included in this report, EPA fully complied with the requirements of the Information Quality Act (also referred to as the Data Quality Act) and EPA's *Guidelines for Ensuring and Maximizing the Quality, Objectivity, Utility, and Integrity of Information Disseminated by the Environmental Protection Agency.*[2]

The process for evaluating indicators is described in more detail below.

A: Identify Candidate Indicators

EPA invited suggestions of new indicators from the public following the release of the April 2010 *Climate Change Indicators in the United States* report, and continues to welcome suggestions at climateindicators@epa.gov. In March 2011, EPA held a meeting of experts on climate change and scientific communication to provide feedback on the first edition of the report. Meeting participants assessed the merits of data in the report and provided suggestions for new content in the future.

Participants suggested a variety of concepts for new indicators and data sources for EPA to consider. These suggestions can be broadly grouped into two categories:

- Additions: Completely new indicators.
- Revisions: Improving an existing indicator by adding or replacing metrics. These revisions would involve obtaining new datasets and vetting their scientific validity.

Suggestions from the participants informed EPA's investigation into candidate indicators for the screening and selection process. As part of this process, existing indicators are re-evaluated as appropriate to ensure they continue to function as intended and they meet EPA's indicator criteria. The process of identifying indicators also includes monitoring the scientific literature, availability of new data, and eliciting expert review.

---

[2] U.S. EPA. 2002. Guidelines for ensuring and maximizing the quality, objectivity, utility, and integrity of information disseminated by the Environmental Protection Agency. EPA/260R-02-008. http://www.epa.gov/quality/informationguidelines/documents/EPA_InfoQualityGuidelines.pdf.

B and C: Research and Screening

Indicator Criteria

EPA screened and selected indicators based on an objective, transparent process that considered the scientific integrity of each proposed indicator, the availability of data, and the value of including the proposed indicator in the report. Each candidate indicator was initially evaluated against fundamental EPA selection criteria to assess whether or not it was reasonable to pursue for inclusion in the upcoming report. These criteria included the peer-review status of the data, the accessibility of the underlying data, the relevance and usefulness of the indicator, its ability to be understood by the public, and its connection to climate change.

Tier 1 Criteria

- Peer-reviewed data
- Feasible to construct
- Usefulness
- Understandable to the public
- Connection to climate change

Tier 2 Criteria

- Transparent, reproducible, and objective
- Broad geographic range
- Actual observations
- Trends over time
- Indicator confidence

The distinction between Tier 1 and Tier 2 criteria is not intended to suggest that the one group is necessarily more important than the other. Rather, EPA determined that a reasonable approach was to consider which criteria must be met before proceeding further and to narrow the list of indicator candidates before the remaining criteria were applied.

Screening Process

EPA researched and screened candidate indicators by creating and populating a database with all of the suggested additions and revisions, then documented the extent to which each of these proposed indicators met each of EPA's criteria. EPA researched and screened in two main stages:

- **Tier 1 screening:** Indicators were scored high, moderate, or low against the Tier 1 criteria. Indicators scoring high or moderate were researched further; indicators scoring low were eliminated from consideration. Many of the suggestions ruled out at this stage were ideas that could lead to viable indicators in the future, but did not yet have any published data.
- **Tier 2 screening:** Indicators deemed appropriate for additional screening were assessed against the Tier 2 criteria. Based on the findings from the complete set of 10 criteria, the indicators were again evaluated and scored high, moderate, or low based on the assessment team's judgment of whether EPA should consider adding them to the report.

Information Sources

To assess each suggested indicator against the criteria, EPA reviewed the scientific literature using numerous methods (including several online databases and search tools) to identify existing data sources and peer-reviewed publications.

In cases where the candidate indicator was not associated with a well-defined metric, EPA conducted a broader survey of the literature to identify the most frequently used metrics. For instance, an indicator related to "community composition" (i.e., biodiversity) was suggested but it was unclear how this variable might best be measured or represented by a metric.

To gather additional information, EPA contacted appropriate subject matter experts, including authors of identified source material, existing data contributors, and collaborators.

D: Indicator Selection

Based on the results of Tier 2 screening, the most promising indicators for the report were developed into proposed indicator summaries. EPA consulted the literature, subject matter experts, and online databases to obtain data for each of these indicators. Upon acquiring sound data and technical documentation, EPA prepared a set of possible graphical mockups for each indicator, along with a summary table that described the proposed metric(s), data sources, limitations, and other relevant information.

Summary information was reviewed by EPA technical staff, and then the indicator concepts that met the screening criteria were formally approved for inclusion in the report.

E: Indicator Development

Approved new and revised indicators were then developed for the inclusion in the report. Graphics, summary text, and technical documentation for all of the proposed new or revised indicators were developed in general accordance with the established format for the original 24 indicators. One priority during development was to make sure each indicator presented its information effectively to a non-technical audience without misrepresenting the underlying source of information.

F: Internal and External Reviews

The complete indicator packages (graphics, summary text, and technical documentation) were subjected to internal review, data provider/collaborator review, and an independent peer review.

Internal Review

The report contents were reviewed at various stages of development in accordance with EPA's standard review protocols for publications. This process included review from EPA technical staff and various levels of management.

Data Provider/Collaborator Review

Organizations and individuals who collected and/or compiled the data (e.g., the National Oceanic and Atmospheric Administration and the U.S. Geological Survey) reviewed the report.

Independent Peer Review

The peer review of the report and technical supporting information followed the procedures in *EPA's Peer Review Handbook*, 3rd Edition (EPA/100/B-06/002)[3] for reports that do not provide influential scientific information. The review was managed by a contractor under the direction of a designated EPA peer review coordinator, who prepared a peer review plan, the scope of work for the review contract, and the charge for the reviewers, but played no role in producing the draft report. Under the terms of the peer review plan, the peer review consisted of 12 experts:

- Two experts in environmental indicators with no specific expertise in climate reviewed the entire report.
- One general expert in the causes and effects of climate change reviewed the entire report.
- Nine subject matter experts each reviewed one chapter. Those experts had the following expertise: greenhouse gas emissions and radiative forcing; climate science and meteorology; measuring sea level; glaciers, ice sheets, and sea ice; heat-related mortality; forests and agriculture; and hydrology.

Two of the 12 reviewers were statisticians; they followed a supplemental charge for statisticians, in addition to the general charge for reviewers.

The peer review charge asked reviewers to provide detailed comments and to indicate whether the report (or chapter) should be published (a) as-is, (b) with changes suggested by the review, (c) only after a substantial revision necessitating a re-review, or (d) not at all. Eight reviewers answered (b). Four answered (c) for at least one chapter, suggesting that alternative methods or data could be used for certain indicators in the draft report. One expert reviewer noted several limitations to the heat-related deaths indicator and suggested significant revisions to the indicator; a full-report reviewer also had similar concerns about this indicator. In total the reviewers provided over 800 comments.

EPA revised the report to address all comments and prepared a spreadsheet to document the response to each comment received. The revised report and EPA's responses were then sent to six of the peer-reviewers for re-review.

Four of the re-reviewers were satisfied with the revised draft. The reviewers of the heat-related deaths indicator provided additional comments and noted that the authors should more clearly articulate certain limitations of this indicator. In response, EPA made further revisions to more clearly document the limitations of the heat-related deaths indicator and address peer-review concerns. The two reviewers reviewed the indicator again and were satisfied with the revisions.

---

[3] U.S. EPA. 2006. Peer review handbook. Third edition. EPA 100/B-06/002.
http://www.epa.gov/peerreview/pdfs/peer_review_handbook_2006.pdf.

G: Periodic Re-Evaluation of Indicators

The process of evaluating indicators includes monitoring the availability of newer data, eliciting expert review, and assessing indicators in light of new science. For example, EPA determined that the underlying methods for developing the Plant Hardiness Zone indicator that appeared in the first edition of *Climate Change Indicators in the United States* (April 2010) had significantly changed, such that updates to the indicator are no longer possible. Thus, EPA removed this indicator from the 2012 edition. Re-evaluation of indicators occurs in the time between EPA publications of the report—about every two to five years.

Summary of Changes to the 2012 Report

The table below highlights major changes made during development of the 2012 version of the report, compared with the 2010 report.

Indicator (number of figures)	Change	Years of data added since 2010 report	Most recent data
U.S. Greenhouse Gas Emissions (3)		2	2010
Global Greenhouse Gas Emissions (3)		3	2008
Atmospheric Concentrations of Greenhouse Gases (4)	Added more halogenated gases	2	2011
Climate Forcing (1)		3	2011
U.S. and Global Temperature (3)		2	2011
High and Low Temperatures (formerly Heat Waves) (4)	Expanded with new metrics	3	2012
U.S. and Global Precipitation (3)		2	2011
Heavy Precipitation (2)		3	2011
Drought (2)	Expanded with new metric	2	2011
Tropical Cyclone Activity (formerly Tropical Cyclone Intensity) (3)	Expanded with new metric	2	2011
Ocean Heat (1)		3	2011
Sea Surface Temperature (1)	New example map	2	2011
Sea Level (2)	New satellite-based data source	3	2011
Ocean Acidity (2)	Replaced map with new metric	7	2012
Arctic Sea Ice (2)	Expanded with new metric	3	2012
Glaciers (2)	New global data source	2	2010
Lake Ice (3)		10	2010
Snowfall (2)	New indicator		2011
Snow Cover (2)	Expanded with new metric	3	2011
Snowpack (1)			2000
Streamflow (3)	New indicator		2009
Ragweed Pollen Season (1)	New indicator		2011
Length of Growing Season (3)		9	2011
Leaf and Bloom Dates (2)	Based on new analytical method	2	2010

Indicator (number of figures)	Change	Years of data added since 2010 report	Most recent data
Bird Wintering Ranges (2)			2005
Heat-Related Deaths (1)	Expanded with new metric and example graphic		2009
Plant Hardiness Zones (2)	Discontinued		2006

Discontinued Indicators

Plant Hardiness Zones; Discontinued in April 2012

Reason for Discontinuation:

This indicator compared the U.S. Department of Agriculture's (USDA's) 1990 Plant Hardiness Zone Map (PHZM) with a 2006 PHZM that the Arbor Day Foundation compiled using similar methods. USDA developed[4] and published a new PHZM in January 2012, reflecting more recent data as well as the use of better analytical methods to delineate zones between weather stations, particularly in areas with complex topography (e.g., many parts of the West). Because of the differences in methods, it is not appropriate to compare the original 1990 PHZM with the new 2012 PHZM to assess change, as many of the apparent zone shifts would reflect improved methods rather than actual temperature change. Further, USDA cautioned users against comparing the 1990 and 2012 PHZMs and attempting to draw any conclusions about climate change from the apparent differences.

For these reasons, EPA chose to discontinue the indicator. EPA will revisit this indicator in the future if USDA releases new editions of the PHZM that allow users to examine changes over time.

For more information about USDA's 2012 PHZM, see: http://planthardiness.ars.usda.gov/PHZMWeb/. The original version of this indicator as it appeared in EPA's 2010 report can be found at: www.epa.gov/climatechange/indicators/download.html.

---

[4] Daly, C., M.P. Widrlechner, M.D., Halbleib, J.I., Smith, and W.P. Gibson. 2012. Development of a new USDA Plant Hardiness Zone Map for the United States. Journal of Applied Meteorology and Climatology 51:242–264.

U.S. Greenhouse Gas Emissions

Identification

1. Indicator Description

This indicator describes emissions of greenhouse gases in the United States and its territories between 1990 and 2010. This indicator reports emissions of greenhouse gases (GHGs) according to their global warming potential, a measure of how much a given amount of the GHG is estimated to contribute to global warming over a selected period of time. For the purposes of comparison, global warming potential values are given in relation to carbon dioxide (CO_2) and are expressed in terms of CO_2 equivalents.

Components of this indicator include:

- U.S. GHG emissions by gas (Figure 1)
- U.S. GHG emissions and sinks by economic sector (Figure 2)
- U.S. GHG emissions per capita and per dollar of GDP (Figure 3)

2. Revision History

April 2010: Indicator posted
December 2011: Updated with data through 2009
April 2012: Updated with data through 2010

Data Sources

3. Data Sources

This indicator uses data and analysis from EPA's *Inventory of U.S. Greenhouse Gas Emissions and Sinks* (U.S. EPA, 2012), an assessment of the anthropogenic sources and sinks of GHGs for the United States and its territories for the period from 1990 to 2010.

4. Data Availability

The complete U.S. GHG inventory is published annually, and the version used to prepare this indicator is publicly available at: www.epa.gov/climatechange/ghgemissions/usinventoryreport.html (U.S. EPA, 2012). The figures in this indicator are taken from the following figures and tables in the inventory report:

- Figure 1 (emissions by gas): Figure ES-1/Table ES-2
- Figure 2 (emissions by economic sector): Figure ES-13/Table ES-7
- Figure 3 (emissions per capita and per dollar gross domestic product [GDP]): Figure ES-15/Table ES-9

The inventory report itself does not present data for the years 1991–1994, 1996–1999, or 2001–2004 due to space constraints. However, data for these years can be obtained by contacting EPA's Climate Change Division (www.epa.gov/climatechange/contactus.html).

Figure 3 includes trends in population and real GDP. EPA obtained population data from the U.S. Census Bureau. These data are publicly available from the Census Bureau's International Data Base at: www.census.gov/population/international/. EPA obtained GDP data from the U.S. Department of Commerce, Bureau of Economic Analysis. These data are publicly available from the Bureau of Economic Analysis website at: www.bea.gov/national/index.htm#gdp.

Methodology

5. Data Collection

This indicator uses data directly from the *Inventory of U.S. Greenhouse Gas Emissions and Sinks* (U.S. EPA, 2012). The inventory presents estimates of emissions derived from direct measurements, aggregated national statistics, and validated models. Specifically, this indicator focuses on the six long-lived greenhouse gases currently covered by agreements under the United Nations Framework Convention on Climate Change (UNFCCC). These compounds are CO_2, methane (CH_4), nitrous oxide (N_2O), selected hydrofluorocarbons (HFCs), selected perfluorocarbons (PFCs), and sulfur hexafluoride (SF_6).

The emission and source activity data used to derive the emission estimates are described thoroughly in EPA's inventory report. The scientifically approved methods can be found in the Intergovernmental Panel on Climate Change's (IPCC's) GHG inventory guidelines (www.ipcc-nggip.iges.or.jp/public/gl/invs1.htm) (IPCC, 2006) and in IPCC's "Good Practice Guidance and Uncertainty Management in National Greenhouse Gas Inventories" (www.ipcc-nggip.iges.or.jp/public/gp/english/) (IPCC, 2000). More discussion of the sampling and data sources associated with the inventory can be found at: www.epa.gov/climatechange/ghgemissions/ .

The U.S. GHG inventory provides a thorough assessment of the anthropogenic emissions by sources and removals by sinks of GHGs for the United States from 1990 to 2010. Although the inventory is intended to be comprehensive, certain identified sources and sinks have been excluded from the estimates (e.g., CO_2 from burning in coal deposits and waste piles, CO_2 from natural gas processing). Sources are excluded from the inventory for various reasons, including data limitations or a lack of thorough understanding of the emission process. The United States is continually working to improve upon the understanding of such sources and seeking to find the data required to estimate related emissions. As such improvements are made, new emission sources are quantified and included in the inventory. For a complete list of excluded sources, see Annex 5 of the U.S. GHG inventory report (www.epa.gov/climatechange/ghgemissions/usinventoryreport.html).

Figure 3 of this indicator compares emission trends with trends in population and U.S. gross domestic product (GDP). Population data were collected by the U.S. Census Bureau. For this indicator, EPA used midyear estimates of the total U.S. population. GDP data were collected by the U.S. Department of Commerce, Bureau of Economic Analysis. For this indicator, EPA used real GDP in chained 2005 dollars, which means the numbers have been adjusted for inflation. See: www.census.gov/population/international/ for the methods used to determine midyear population estimates for the United States. See: www.bea.gov/methodologies/index.htm#national_meth for the methods used to determine GDP.

6. Indicator Derivation

The U.S. GHG inventory was constructed following scientific methods that can be found in the Intergovernmental Panel on Climate Change's (IPCC's) *Guidelines for National Greenhouse Gas Inventories* (IPCC, 2006) and in IPCC's *Good Practice Guidance and Uncertainty Management in National Greenhouse Gas Inventories* (IPCC, 2000). EPA's annual inventory reports and IPCC's inventory development guidelines have been extensively peer reviewed and are widely viewed as providing scientifically sound representations of GHG emissions.

The U.S. GHG inventory is not based on a specific sampling plan or analytical procedures per se. However, U.S. EPA (2012) provides a complete description of methods and data sources that allowed EPA to calculate GHG emissions for the various industrial sectors and source categories. Further information on the inventory design can be obtained by contacting EPA's Climate Change Division (www.epa.gov/climatechange/contactus.html).

The inventory covers U.S. data for the years 1990 to 2010, and no attempt has been made to incorporate other locations or project data forward or backward from this time window. Some degree of extrapolation and interpolation was needed to develop comprehensive estimates of emissions in a few sectors and sink categories, but in most cases, observations and estimates from the year in question were sufficient to generate the necessary data points.

This indicator reports trends exactly as they appear in EPA's GHG inventory (U.S. EPA, 2012). The indicator presents emission data in units of million metric tons of CO_2 equivalents, which are conventionally used in GHG inventories prepared worldwide because they adjust for the various global warming potentials of different gases. This analysis reflects the use of 100-year global warming potentials.

Figure 1.U.S. Greenhouse Gas Emissions by Gas, 1990–2010

EPA plotted total emissions for each gas, not including the influence of sinks, which would be difficult to interpret in a breakdown by gas. EPA combined the emissions of HFCs, PFCs, and SF_6 into a single category so the magnitude of these emissions would be visible in the graph.

Figure 2. U.S. Greenhouse Gas Emissions and Sinks by Economic Sector, 1990–2010

EPA converted a line graph in the original inventory report (U.S. EPA, 2012) into a stacked area graph showing emissions by economic sector. U.S. territories are treated as a separate sector in the inventory report, and because territories are not an economic sector in the truest sense of the word, they have been excluded from this part of the indicator. Unlike Figure 1, Figure 2 includes sinks below the x-axis.

Figure 3. U.S. Greenhouse Gas Emissions per Capita and per Dollar of GDP, 1990–2010

EPA determined emissions per capita and emissions per unit of real GDP using simple division. In order to show all four trends (population, GDP, emissions per capita, and emissions per unit GDP) on the same scale, EPA normalized each trend to an index value of 100 for the year 1990.

7. Quality Assurance and Quality Control

Quality assurance and quality control (QA/QC) have always been an integral part of the U.S. national system for inventory development. EPA and its partner agencies have implemented a systematic approach to QA/QC for the annual U.S. GHG inventory, following procedures that have been formalized in accordance with a QA/QC plan and the UNFCCC reporting guidelines. Those interested in documentation of the various QA/QC procedures should send such queries to EPA's Climate Change Division (www.epa.gov/climatechange/contactus.html).

Analysis

8. Comparability Over Time and Space

The GHG emissions data presented in this indicator are viewed as being highly comparable over time and space.

9. Sources of Uncertainty

Some estimates, such as those for CO_2 emissions from energy-related activities and cement processing, are considered to have low uncertainties. For some other categories of emissions, however, a lack of data or an incomplete understanding of how emissions are generated increases the uncertainty associated with the estimates presented.

Recognizing the benefit of conducting an uncertainty analysis, the UNFCCC reporting guidelines follow the recommendations of IPCC (2000) and require that countries provide single point uncertainty estimates for many sources and sink categories. The U.S. GHG inventory (U.S. EPA, 2012) provides a qualitative discussion of uncertainty for all sources and sink categories, including specific factors affecting the uncertainty surrounding the estimates. Most sources also have a quantitative uncertainty assessment in accordance with the new UNFCCC reporting guidelines. Thorough discussion of these points can be found in U.S. EPA (2012). Note that Annex 7 of the inventory publication is devoted entirely to uncertainty in the inventory estimates.

For a general idea of the degree of uncertainty in U.S. emission estimates, WRI (2011) provides the following information: "Using IPCC Tier 2 uncertainty estimation methods, EIA (2002) estimated uncertainties surrounding a simulated mean of CO_2 (-1.4% to 1.3%), CH_4 (-15.6% to 16%), and N_2O (-53.5% to 54.2%). Uncertainty bands appear smaller when expressed as percentages of total estimated emissions: CO_2 (-0.6% to 1.7%), CH_4 (-0.3% to 3.4%), and N_2O (-1.9% to 6.3%)."

Overall, these sources of uncertainty are not expected to have a considerable impact on this indicator's conclusions. Even considering the uncertainties of omitted sources and lack of precision in known and estimated sources, this indicator provides a generally accurate picture of aggregate trends in GHG emissions over time, and hence the overall conclusions inferred from the data are solid. The U.S. GHG inventory represents the most comprehensive and reliable data set available to characterize GHG emissions in the United States.

10. Sources of Variability

Within each sector (e.g., electricity generation), GHG emissions can vary considerably across the individual point sources, and many factors contribute to this variability (e.g., different production levels, fuel type, air pollution controls). EPA's inventory methods account for this variability among individual emission sources.

11. Statistical/Trend Analysis

This indicator presents a time series of national emissions estimates. No special statistical techniques or analyses were used to characterize the long-term trends or their statistical significance.

12. Data Limitations

Factors that may impact the confidence, application, or conclusions drawn from this indicator are as follows:

1. This indicator does not yet include emissions of GHGs or other radiatively important substances that are not explicitly covered by the UNFCCC and its subsidiary protocol. Thus, it excludes such gases as those controlled by the Montreal Protocol and its Amendments, including chlorofluorocarbons and hydrochlorofluorocarbons. Although the United States reports the emissions of these substances as part of the U.S. GHG inventory (see Annex 6.2 of U.S. EPA [2012]), the origin of the estimates is fundamentally different from those of the other GHGs, and therefore these emissions cannot be compared directly with the other emissions discussed in this indicator.
2. This indicator does not include aerosols and other emissions that affect radiative forcing and that are not well-mixed in the atmosphere, such as sulfate, ammonia, black carbon, and organic carbon. Emissions of these compounds are highly uncertain and have qualitatively different effects from the six types of emissions in this indicator.
3. This indicator does not include emissions of other compounds—such as carbon monoxide, nitrogen oxides, nonmethane volatile organic compounds, and substances that deplete the stratospheric ozone layer—that indirectly affect the Earth's radiative balance (for example, by altering GHG concentrations, changing the reflectivity of clouds, or changing the distribution of heat fluxes).
4. The U.S. GHG inventory does not account for "natural" emissions of GHGs from sources such as wetlands, tundra soils, termites, and volcanoes. These excluded sources are discussed in Annex 5 of the U.S. GHG inventory (U.S. EPA, 2012). The "land use," "land-use change," and "forestry" categories in U.S. EPA (2012) do include emissions from changes in the forest inventory due to fires, harvesting, and other activities, as well as emissions from agricultural soils.

References

IPCC (Intergovernmental Panel on Climate Change). 2000. Good practice guidance and uncertainty management in national greenhouse gas inventories. www.ipcc-nggip.iges.or.jp/public/gp/english.

IPCC (Intergovernmental Panel on Climate Change). 2006. IPCC guidelines for national greenhouse gas inventories. www.ipcc-nggip.iges.or.jp/public/2006gl/index.html.

U.S. EPA. 2012. Inventory of U.S. greenhouse gas emissions and sinks: 1990–2010. www.epa.gov/climatechange/ghgemissions/usinventoryreport.html.

WRI (World Resources Institute). 2011. CAIT: Greenhouse gas sources & methods. http://cait.wri.org/downloads/cait_ghgs.pdf.

Global Greenhouse Gas Emissions

Identification

1. Indicator Description

This indicator describes emissions of greenhouse gases (GHGs) worldwide since 1990. Tracking GHG emissions worldwide provides a context for understanding the United States' role in addressing climate change.

Components of this indicator include:

- Global GHG emissions by gas (Figure 1)
- Global GHG emissions by sector (Figure 2)
- Global GHG emissions by regions of the world (Figure 3)

2. Revision History

April 2010: Indicator posted
December 2011: Updated with new and revised data points
April 2012: Updated with revised data points

Data Sources

3. Data Sources

This indicator is based on data from the World Resources Institute's (WRI's) Climate Analysis Indicators Tool (CAIT), a database of anthropogenic sources and sinks of GHGs worldwide. CAIT has compiled data from a variety of GHG emission inventories. In general, a GHG emission inventory consists of estimates derived from direct measurements, aggregated national statistics, and validated models. EPA obtained data from CAIT Version 9.0.

4. Data Availability

All indicator data can be obtained from the WRI CAIT database at: http://cait.wri.org. These data are available to the public, but users must register (at no charge) to receive full access. CAIT includes documentation that describes the various data fields and their sources.

CAIT compiles data from a variety of other databases and inventories, including products from EPA, the U.S. Carbon Dioxide Information Analysis Center (CDIAC), and the International Energy Agency. Many of these original data sources are publicly available. For information on all the sources used to populate the CAIT database, see WRI (2011a). For a list of data sources by country, by gas, and by source or sink category, see: http://cait.wri.org/cait.php?page=notes&chapt=2. Data for this particular indicator were compiled by WRI largely from the following sources:

- Boden et al. (2011)

- EIA (2011)
- European Commission et al. (2009)
- IEA (2010)
- U.S. EPA (2006)

The figures in this indicator are taken from the following reports within CAIT:

- Figure 1 (emissions by gas): "Compare Gases" analysis.
- Figure 2 (emissions by sector): "Compare Sectors" analysis.
- Figure 3 (carbon dioxide [CO_2] emissions by region): "GHG Emissions" indicator → "Yearly Emissions" (customize "Countries & Regions" to display data by continent). See: http://cait.wri.org/cait.php?page=notes&chapt=3 for a listing of which countries belong to each region. Note that EPA combined a few regions as described in Section 6.

Methodology

5. Data Collection

This indicator focuses on emissions of the six compounds or groups of compounds currently covered by agreements under the United Nations Framework Convention on Climate Change (UNFCCC). These compounds are CO_2, methane (CH_4), nitrous oxide (N_2O), selected hydrofluorocarbons (HFCs), selected perfluorocarbons (PFCs), and sulfur hexafluoride (SF_6). This indicator presents emission data in units of million metric tons of CO_2 equivalents. These units are conventionally used in GHG inventories prepared worldwide because they adjust for the different global warming potentials of different gases.

The data originally come from a variety of GHG inventories. Some have been prepared by national governments; others by international agencies. Data collection techniques (e.g., survey design) vary depending on the source or parameter. Although the CAIT database is intended to be comprehensive, the organizations that develop inventories are continually working to improve their understanding of emission sources and how best to quantify them.

Inventories often use some degree of extrapolation and interpolation to develop comprehensive estimates of emissions in a few sectors and sink categories, but in most cases, observations and estimates from the year in question were sufficient to generate the necessary data points.

GHG inventories are not based on specific sampling plans *per se*. However, documents are available that describe how most inventories have been constructed. For example, U.S. EPA (2012) describes all the procedures used to estimate GHG emissions for EPA's annual U.S. inventory. See the Intergovernmental Panel on Climate Change's (IPCC's) GHG inventory guidelines (IPCC, 2006) and IPCC's *Good Practice Guidance and Uncertainty Management in National Greenhouse Gas Inventories* (IPCC, 2000) for additional guidance that many countries and organizations follow when constructing GHG inventories.

6. Indicator Derivation

This indicator reports selected metrics from WRI's CAIT database, which compiles data from the most reputable GHG inventories around the world. WRI (2011b) provides an overview of how the CAIT database was constructed, and WRI (2011a) describes the data sources and methods used to populate the database. WRI's role is largely to assemble data from other sources, all of which have been critically reviewed. As a result, the totals reported in CAIT are consistent with other compilations such as a European tool called the Emissions Database for Global Atmospheric Research (EDGAR) (http://edgar.jrc.ec.europa.eu/index.php), which has been cited in reports by IPCC. EDGAR and CAIT share many of the same underlying data sources.

The most comprehensive estimates are available beginning in 1990. Global emission estimates for CO_2 are available annually through 2008, while global estimates for gases other than CO_2 are available only at five-year intervals through 2005. Thus, Figures 1 and 2 (which show all GHGs) plot values for 1990, 1995, 2000, and 2005. WRI and EPA made no attempt to interpolate estimates for the interim years.

All three figures in this indicator include emissions due to international transport (i.e., aviation and maritime bunker fuel). These emissions are not included in the U.S. Greenhouse Gas Emissions indicator because they are international by nature, and not necessarily reflected in individual countries' emission inventories.

The three figures do not include estimates of emissions associated with LULUCF. Although global estimates are available for this sector, different estimates vary widely, and they have much greater uncertainties than many of the other sectors that this indicator covers.

The indicator presents emission data in units of million metric tons of CO_2 equivalents, which are conventionally used in GHG inventories prepared worldwide because they adjust for the various global warming potentials of different gases. This analysis reflects the use of 100-year global warming potentials.

Figure 1. Global Greenhouse Gas Emissions by Gas, 1990–2005

EPA plotted total emissions for each gas, combining the emissions of HFCs, PFCs, and SF_6 into a single category so the magnitude of these emissions would be visible in the graph. EPA formatted the graph as a series of stacked columns instead of a continuous stacked area because complete estimates for all gases are available only every five years, and it would be misleading to suggest that information is known about trends in the interim years.

Figure 2. Global Greenhouse Gas Emissions by Sector, 1990–2005

EPA plotted total GHG emissions by IPCC sector. IPCC sectors are different from the sectors used in Figure 2 of the U.S. Greenhouse Gas Emissions indicator, which uses an economic sector breakdown that is not available on a global scale. EPA formatted the graph as a series of stacked columns instead of a continuous stacked area because complete estimates for all gases are available only every five years, and it would be misleading to suggest that information is known about trends in the interim years.

Figure 3. Global Carbon Dioxide Emissions by Region, 1990–2008

In order to show data at more than four points in time, EPA elected to display emissions by region for CO_2 only, as CO_2 emission estimates are available with annual resolution. EPA performed simple math to ensure that no emissions were double-counted across the regions. Specifically, EPA subtracted U.S. totals from North American totals, leaving "Other North America" as a separate category. EPA combined a few regions for the graphic: "Other North America" also includes Central America and the Caribbean, and "Africa and the Middle East" includes North Africa as well as Sub-Saharan Africa.

Indicator Development

In the course of developing and revising this indicator, EPA considered data from a variety of sources, including WRI's CAIT (http://cait.wri.org) and EDGAR (http://edgar.jrc.ec.europa.eu/index.php). EPA compared data obtained from CAIT and EDGAR for global carbon dioxide emissions, and found the two data sources were highly comparable (differences of less than 2 percent for all years). EPA also compared emissions associated with land use, land-use change, and forestry (LULUCF) from both CAIT and EDGAR, and found much larger differences that depended on the source and how it characterized certain types of LULUCF emissions. Because of these differences and the larger uncertainties associated with the LULUCF sector on a global scale, EPA chose not to include the LULUCF sector in any of the figures for this indicator.

For the purposes of CAIT, WRI essentially serves as a secondary compiler of global emissions data, drawing on internationally recognized inventories from government agencies and using extensively peer-reviewed datasets. EPA has determined that WRI does not perform additional interpolations on the data but rather makes certain basic decisions in order to allocate emissions to certain countries (e.g., in the case of historical emissions from Soviet republics). These methods are described in CAIT's supporting documentation, which EPA carefully reviewed to assure the credibility of the source.

7. Quality Assurance and Quality Control

Quality assurance and quality control (QA/QC) documentation is not explicitly provided with the full CAIT database, but many of the contributing sources have documented their QA/QC procedures. For example, EPA and its partner agencies have implemented a systematic approach to QA/QC for the annual U.S. GHG inventory, following procedures that have recently been formalized in accordance with a QA/QC plan and the UNFCCC reporting guidelines. Those interested in documentation of the various QA/QC procedures for the U.S. inventory should send such queries to EPA's Climate Change Division (www.epa.gov/climatechange/contactus.html). QA/QC procedures for other sources can generally be found in the documentation that accompanies the sources cited in Section 4.

Analysis

8. Comparability Over Time and Space

Some inventories have been prepared by national governments; others by international agencies. Data collection techniques (e.g., survey design) vary depending on the source or parameter. To the extent possible, inventories follow a consistent set of best practice guidelines described in IPCC (2000, 2006).

9. Sources of Uncertainty

In general, all emission estimates will have some inherent uncertainty. Estimates of CO_2 emissions from energy-related activities and cement processing are often considered to have the lowest uncertainties, but even these data can have errors as a result of uncertainties in the numbers from which they are derived, such as national energy use data. In contrast, estimates of emissions associated with land use, land-use change, and forestry may have particularly large uncertainties, which is why this sector has been excluded from EPA's indicator. Uncertainties are generally larger for non-CO_2 gases.

WRI (2011a) provides the following information about uncertainty in U.S. emissions: "Using IPCC Tier 2 uncertainty estimation methods, EIA (2002) estimated uncertainties surrounding a simulated mean of CO_2 (-1.4% to 1.3%), CH_4 (-15.6% to 16%), and N_2O (-53.5% to 54.2%). Uncertainty bands appear smaller when expressed as percentages of total estimated emissions: CO_2 (-0.6% to 1.7%), CH_4 (-0.3% to 3.4%), and N_2O (-1.9% to 6.3%)."

Uncertainties are expected to be greater in developing countries, due in some cases to varying quality of underlying activity data and uncertain emission factors (WRI, 2011a).

For specific information about uncertainty, users should refer to documentation from the individual data sources cited in Section 4. Uncertainty estimates are available from the underlying national inventories in some cases, in part because the UNFCCC reporting guidelines follow the recommendations of IPCC (2000) and require countries to provide single point uncertainty estimates for many sources and sink categories. For example, the U.S. GHG inventory (U.S. EPA, 2012) provides a qualitative discussion of uncertainty for all sources and sink categories, including specific factors affecting the uncertainty surrounding the estimates. Most sources also have a quantitative uncertainty assessment in accordance with the new UNFCCC reporting guidelines. Thorough discussion of these points can be found in U.S. EPA (2012). Note that Annex 7 of EPA's inventory publication is devoted entirely to uncertainty in the inventory estimates.

Uncertainty is not expected to have a considerable impact on this indicator's conclusions. Uncertainty is indeed present in all emission estimates, in some cases to a great degree—especially for non-CO_2 gases in developing countries. At an aggregate global scale, however, this indicator accurately depicts the overall direction and magnitude of GHG emission trends over time, and hence the overall conclusions inferred from the data are solid.

10. Sources of Variability

On a national or global scale, year-to-year variability in GHG emissions can arise from a variety of factors such as economic conditions, fuel prices, and government actions. Overall, variability is not expected to have a considerable impact on this indicator's conclusions.

11. Statistical/Trend Analysis

This indicator does not report on the slope of the apparent trends in global GHG emissions, nor does it calculate the statistical significance of these trends. The "Key Points" describe percentage change between 1990 and the most recent year of data—an endpoint-to-endpoint comparison, not necessarily a trend line of best fit.

12. Data Limitations

Factors that may impact the confidence, application, or conclusions drawn from this indicator are as follows:

1. This indicator does not yet include emissions of GHGs or other radiatively important substances that are not explicitly covered by the UNFCCC and its subsidiary protocol. Thus, it excludes such gases as those controlled by the Montreal Protocol and its Amendments, including chlorofluorocarbons and hydrochlorofluorocarbons. Although some countries report emissions of these substances, the origin of the estimates is fundamentally different from those of other GHGs, and therefore these emissions cannot be compared directly with the other emissions discussed in this indicator.
5. This indicator does not include aerosols and other emissions that affect radiative forcing and that are not well-mixed in the atmosphere, such as sulfate, ammonia, black carbon, and organic carbon. Emissions of these compounds are highly uncertain and have qualitatively different effects from the six types of emissions in this indicator.
6. This indicator does not include emissions of other compounds—such as carbon monoxide, nitrogen oxides, nonmethane volatile organic compounds, and substances that deplete the stratospheric ozone layer—which indirectly affect the Earth's radiative balance (for example, by altering GHG concentrations, changing the reflectivity of clouds, or changing the distribution of heat fluxes).
7. This indicator does not account for emissions associated with land use, land-use change, and forestry.
8. This indicator does not account for "natural" emissions of GHGs, such as from wetlands, tundra soils, termites, and volcanoes.
9. Global emission data for non-CO_2 GHGs are available only at five-year intervals. Thus, Figures 1 and 2 show data for only four points in time: 1990, 1995, 2000, and 2005.

References

Boden, T.A., G. Marland, and R. J. Andres. 2011. Global, regional, and national fossil fuel CO_2 emissions. Carbon Dioxide Information Analysis Center, Oak Ridge National Laboratory, U.S. Department of Energy. http://cdiac.ornl.gov/trends/emis/overview_2008.html.

EIA (U.S. Energy Information Administration). 2011. International energy statistics. www.eia.gov/cfapps/ipdbproject/IEDIndex3.cfm?tid=90&pid=44&aid=8 .

European Commission, Joint Research Centre/Netherlands Environmental Assessment Agency. 2009. Emission Database for Global Atmospheric Research (EDGAR), version 4.0. http://edgar.jrc.ec.europa.eu/index.php.

IEA (International Energy Agency). 2010. CO_2 emissions from fuel combustion (2010 edition). http://data.iea.org/ieastore/statslisting.asp.

IPCC (Intergovernmental Panel on Climate Change). 2000. Good practice guidance and uncertainty management in national greenhouse gas inventories. www.ipcc-nggip.iges.or.jp/public/gp/english.

IPCC (Intergovernmental Panel on Climate Change). 2006. IPCC guidelines for national greenhouse gas inventories. www.ipcc-nggip.iges.or.jp/public/2006gl/index.html.

U.S. EPA. 2006. Global anthropogenic non-CO_2 greenhouse gas emissions: 1990–2020. www.epa.gov/climatechange/EPAactivities/economics/nonco2projections.html.

U.S. EPA. 2012. Inventory of U.S. greenhouse gas emissions and sinks: 1990–2010.
www.epa.gov/climatechange/ghgemissions/usinventoryreport.html.

WRI (World Resources Institute). 2011a. CAIT: Greenhouse gas sources & methods.
http://cait.wri.org/downloads/cait_ghgs.pdf.

WRI (World Resources Institute). 2011b. CAIT: Indicator framework paper.
http://cait.wri.org/downloads/framework_paper.pdf.

Atmospheric Concentrations of Greenhouse Gases

Identification

1. Indicator Description

This indicator describes how the levels of major greenhouse gases (GHGs) in the atmosphere have changed over geological time and in recent years. Changes in atmospheric GHGs, in part caused by human activities, affect the amount of energy held in the Earth-atmosphere system and thus affect the Earth's climate.

Components of this indicator include:

- Global atmospheric concentrations of carbon dioxide over time (Figure 1)
- Global atmospheric concentrations of methane over time (Figure 2)
- Global atmospheric concentrations of nitrous oxide over time(Figure 3)
- Global atmospheric concentrations of selected halogenated gases over time (Figure 4)

2. Revision History

April 2010: Indicator posted
December 2011: Updated with data through 2010
May 2012: Updated with data through 2011
July 2012: Added nitrogen trifluoride to Figure 4

Data Sources

3. Data Sources

Ambient concentration data used to develop this indicator were taken from the following sources:

Figure 1. Global Atmospheric Concentrations of Carbon Dioxide Over Time

- EPICA Dome C, Antarctica: approximately 647,426 BC to 411,548 BC—Siegenthaler et al. (2005)
- Vostok Station, Antarctica: approximately 415,157 BC to 339 BC—Barnola et al. (2003)
- EPICA Dome C, Antarctica: approximately 9002 BC to 1515 AD—Flückiger et al. (2002)
- Law Dome, Antarctica, 75-year smoothed: approximately 1010 AD to 1975 AD—Etheridge et al. (1998)
- Siple Station, Antarctica: approximately 1744 AD to 1953 AD—Neftel et al. (1994)
- Mauna Loa, Hawaii: 1959 AD to 2011 AD—NOAA (2012a)
- Barrow, Alaska: 1974 AD to 2011 AD; Cape Matatula, American Samoa: 1976 AD to 2011 AD; South Pole, Antarctica: 1976 AD to 2011 AD—NOAA (2012c)
- Cape Grim, Australia: 1992 AD to 2006 AD; Shetland Islands, Scotland: 1993 AD to 2002 AD—Steele et al. (2007)

- Lampedusa Island, Italy: 1993 AD to 2000 AD—Chamard et al. (2001)

Figure 2. Global Atmospheric Concentrations of Methane Over Time

- EPICA Dome C, Antarctica: approximately 646,729 BC to 1888 AD—Spahni et al. (2005)
- Vostok Station, Antarctica: approximately 415,172 BC to 346 BC—Petit et al. (1999)
- Greenland GISP2 ice core: approximately 87,798 BC to 8187 BC; Byrd Station, Antarctica: approximately 85,929 BC to 6748 BC; Greenland GRIP ice core: approximately 46,933 BC to 8129 BC—Blunier and Brook (2001)
- EPICA Dome C, Antarctica: approximately 8945 BC to 1760 AD—Flückiger et al. (2002)
- Law Dome, Antarctica: approximately 1008 AD to 1980 AD; various Greenland locations: approximately 1075 AD to 1885 AD—Etheridge et al. (2002)
- Greenland Site J: approximately 1598 AD to 1951 AD—WDCGG (2005)
- Cape Grim, Australia: 1984 AD to 2010 AD—NOAA (2011b)
- Mauna Loa, Hawaii: 1987 AD to 2011 AD—NOAA (2012b)
- Shetland Islands, Scotland: 1993 AD to 2001 AD—Steele et al. (2002)

Figure 3. Global Atmospheric Concentrations of Nitrous Oxide Over Time

- Greenland GISP2 ice core: approximately 104,301 BC to 1871 AD; Taylor Dome, Antarctica: approximately 30,697 BC to 497 BC—Sowers et al. (2003)
- EPICA Dome C, Antarctica: approximately 9000 BC to 1780 AD—Flückiger et al. (2002)
- Antarctica: approximately 1756 AD to 1964 AD—Machida et al. (1995)
- Antarctica: approximately 1903 AD to 1976 AD—Battle et al. (1996)
- Cape Grim, Australia: 1979 AD to 2010 AD—AGAGE (2012)
- South Pole, Antarctica: 1998 AD to 2011 AD; Barrow, Alaska: 1999 AD to 2011 AD; Mauna Loa, Hawaii: 2000 AD to 2011 AD—NOAA (2012d)

Figure 4. Global Atmospheric Concentrations of Selected Halogenated Gases, 1978–2010

Global average atmospheric concentration data for selected halogenated gases were obtained from the following sources:

- National Oceanic and Atmospheric Administration (NOAA, 2011a) for halon-1211.
- Weiss et al. (2008) and Arnold et al. (2012) for nitrogen trifluoride.
- Advanced Global Atmospheric Gases Experiment (AGAGE, 2011) for all other species shown.

A similar figure based on AGAGE data appeared in the Intergovernmental Panel on Climate Change's (IPCC's) Fourth Assessment Report (see Figure 2.6 in IPCC, 2007).

4. Data Availability

The data used to develop Figures 1, 2, and 3 of this indicator are publicly available and can be accessed from the references and websites listed in Section 3. There are no known confidentiality issues.

Data for Figure 4 were provided in spreadsheet form by Dr. Ray Wang of the AGAGE project team and Dr. Stephen Montzka of NOAA. AGAGE and NOAA websites (http://agage.eas.gatech.edu/ and www.esrl.noaa.gov/gmd/hats/) provide access to underlying station-specific data and selected averages, but not all of the global averages that are shown in Figure 4. Nitrogen trifluoride data are based on measurements that were originally published by Weiss et al. (2008), an additional set of 2011 measurements published in Arnold et al. (2012), and a correction factor in Arnold et al. (2012) that EPA applied to the earlier data.

Methodology

5. Data Collection

This indicator shows trends in atmospheric concentrations of several major GHGs that enter the atmosphere at least in part because of human activities: carbon dioxide (CO_2), methane (CH_4), nitrous oxide (N_2O), and selected halogenated gases. This indicator aggregates comparable, high-quality data from individual studies that each focused on different locations and time frames. Recent data (since the mid-20th century) come from global networks that use standard monitoring techniques to measure the concentrations of gases in the atmosphere. Older data come from ice cores—specifically, measurements of gas concentrations in air bubbles that were trapped in ice at the time the ice was formed. Scientists have spent years developing and refining methods of measuring gases in ice cores as well as methods of dating the corresponding layers of ice to determine their age. Ice core measurements are a widely used method of reconstructing the composition of the atmosphere before the advent of direct monitoring techniques.

This indicator presents a compilation of data generated by numerous sampling programs. The citations listed in Section 3 describe the specific approaches taken by each program. Gases are measured by mole fraction relative to dry air.

Most of the GHGs presented in this indicator are considered to be well-mixed globally, due in large part to their long residence times in the atmosphere. Thus, while measurements over geological time tend to be available only for regions where ice cores can be collected (e.g., the Arctic and Antarctic regions), these measurements are believed to adequately represent concentrations worldwide. Recent monitoring data have been collected from a greater variety of locations, and the results show that concentrations and trends are indeed very similar throughout the world, although relatively small variations can be apparent across different locations.

Most of the gases shown in Figure 4 have been measured around the world numerous times per year. One exception is nitrogen trifluoride, for which measurements are not yet widespread. The curve for nitrogen trifluoride in Figure 4 is based on measurements of six air samples collected at Trinidad Head, California, between 1998 and 2008, and a series of measurements at La Jolla, California. in 2011. Measurements of air samples collected before 1998 have also been made, but they are not included in this figure because of larger gaps in time between measurements. Northern Hemisphere concentrations of this gas are expected to be slightly higher than the global average because of the distribution of sources—particularly the electronics industry.

Nitrogen trifluoride was measured via the Medusa gas chromatography with mass spectrometry (GCMS) system, with refinements described in Weiss et al. (2008) and Arnold et al. (2012). Mole fractions of the other halogenated gases were collected via AGAGE's Medusa GCMS system or similar methods employed by NOAA.

6. Indicator Derivation

EPA obtained and compiled data from the various GHG measurement programs and plotted these data in graphs. Figures 1, 2, and 3 plot data at annual or lower resolution; with ice cores, consecutive data points are often spaced many years apart. Figure 4 plots data at sub-annual intervals. EPA used the data exactly as reported by the organizations that collected them, with the following exceptions:

- Some of the recent time series for CO_2, CH_4, and N_2O consisted of monthly measurements. EPA averaged these monthly measurements to arrive at annual values to plot in the graphs. A few years did not have data for all 12 months. If at least nine months of data were present in a given year, EPA averaged the available data to arrive at an annual value. If fewer than nine monthly measurements were available, that year was excluded from the graph.
- Some ice core records were reported in terms of the age of the sample or the number of years before present. EPA converted these dates into calendar years.
- A few ice core records had multiple values at the same point in time (i.e., two or more different measurements for the same year). These values were generally comparable and never varied by more than 4.8 percent. In such cases, EPA averaged the values to arrive at a single atmospheric concentration per year.
- Although measurements have been made of nitrogen trifluoride in air samples collected before 1998, EPA elected to start the nitrogen trifluoride time series at 1998 because of large time gaps between measurements prior to 1998.

Figures 1, 2, and 3 label each trend line according to the location where measurements were collected. No methods were used to portray data for locations other than where measurements were made. However, the indicator does imply that the values in the graphs represent global atmospheric concentrations—an appropriate assumption because the gases covered by this indicator have long residence times in the atmosphere and are considered to be well-mixed. In the indicator text, the Key Points refer to the concentration for the most recent year available. If data were available for more than one location, the text refers to the average concentration across these locations.

Figure 4 presents one trend line for each halogenated gas, and these lines represent average concentrations across all measurement sites (typically worldwide, except for nitrogen trifluoride as noted in Section 5). These data represent monthly average mole fractions for each species, except for nitrogen trifluoride, which relies on a smaller number of individual samples.

Data are available for additional halogenated species, but to make the most efficient use of the space available, EPA selected a subset of gases that are relatively common, have several years of data available, show marked growth trends (either positive or negative), and/or collectively represent most of the major categories of halogenated gases. The inclusion of nitrogen trifluoride here is based on several factors. Although nitrogen trifluoride has relatively fewer measurements available, the data are representative of atmospheric concentrations in the Northern Hemisphere. Like perfluoromethane (PFC-14 or CF_4), perfluoroethane (PFC-116 or C_2F_6), and sulfur hexafluoride, nitrogen trifluoride is a widely produced, fully fluorinated gas with a very high 100-year global warming potential (17,200) and a long atmospheric lifetime (740 years). Nitrogen trifluoride has experienced a rapid increase in emissions (i.e., more than 10 percent per year) due to its use in manufacturing semiconductors, flat screen displays, and thin film solar cells. It began to replace perfluoroethane in the electronics industry in the late 1990s.

To examine the possible influence of phase-out and substitution activities under the Montreal Protocol on Substances That Deplete the Ozone Layer, EPA divided Figure 4 into two panels: one for substances officially designated as "ozone-depleting" and one for all other halogenated gases.

No attempt was made to project concentrations backward before the beginning of the ice core record (or the start of monitoring, in the case of Figure 4) or forward into the future.

7. Quality Assurance and Quality Control

The data for this indicator have generally been taken from carefully constructed, peer-reviewed studies. Quality assurance and quality control procedures are addressed in the individual studies, which are cited in Section 3. Additional documentation of these procedures can be obtained by consulting with the principal investigators who developed each of the data sets.

Analysis

8. Comparability Over Time and Space

Data have been collected using a variety of methods over time and space. However, these methodological differences are expected to have little bearing on the overall conclusions for this indicator. The concordance of trends among multiple data sets collected using different program designs provides some assurance that the trends depicted actually represent atmospheric conditions, rather than some artifact of sampling design.

The gases covered in this indicator are all long-lived GHGs that are relatively evenly distributed globally. Thus, measurements collected at one particular location have been shown to be representative of average concentrations worldwide.

9. Sources of Uncertainty

Direct measurements of atmospheric concentrations, which cover approximately the last 50 years, are of a known and high quality. Generally, standard errors and accuracy measurements are computed for the data.

For ice core measurements, uncertainties result from the actual gas measurements as well as the dating of each sample. Uncertainties associated with the measurements are believed to be relatively small, although diffusion of gases from the samples might also add to the measurement uncertainty. Dating accuracy for the ice cores is believed to be within plus or minus 20 years, depending on the method used and the time period of the sample. However, this level of uncertainty is insignificant when considering that some ice cores characterize atmospheric conditions for time frames more than 100,000 years ago. The original scientific publications (see Section 3) provide more detailed information on the estimated uncertainty within the individual data sets.

Visit the Carbon Dioxide Information Analysis Center (CDIAC) website (http://cdiac.esd.ornl.gov/by_new/bysubjec.html#atmospheric) for more information on the accuracy of both direct and ice core measurements.

Overall, the concentration increase in GHGs in the past century is far greater than the estimated uncertainty of the underlying measurement methodologies. Otherwise stated, it is highly unlikely that the concentration trends depicted in this indicator are artifacts of uncertainty.

10. Sources of Variability

Atmospheric concentrations of GHGs vary with both time and space. However, the data on atmospheric GHG concentrations have extraordinary temporal coverage. For carbon dioxide, methane, and nitrous oxide, concentration data span several hundred thousand years; and for the halogenated gases, data span virtually the entire period during which these largely synthetic gases were widely used. While spatial coverage of monitoring stations is more limited, most of the GHGs presented in this indicator are considered to be well-mixed globally, due in large part to their long residence times in the atmosphere.

11. Statistical/Trend Analysis

This indicator presents a time series of atmospheric concentrations of GHGs. No statistical techniques or analyses were used to characterize the long-term trends or their statistical significance.

12. Data Limitations

Factors that may impact the confidence, application, or conclusions drawn from this indicator are as follows:

1. This indicator does not track water vapor because of its spatial and temporal variability. Human activities have only a small direct impact on water vapor concentrations, but there are indications that increasing global temperatures are leading to increasing levels of atmospheric humidity (Dai et al., 2011).
2. Some radiatively important atmospheric constituents that are substantially affected by human activities (such as tropospheric ozone, black carbon, aerosols, and sulfates) are not included in this indicator because of their spatial and temporal variability.
3. Ice core measurements are not taken in real time, which introduces some error into the date of the sample. Dating accuracy for the ice cores ranges up to plus or minus 20 years (often less), depending on the method used and the time period of the sample. Diffusion of gases from the samples, which would tend to reduce the measured values, could also add a small amount of uncertainty.

References

AGAGE (Advanced Global Atmospheric Gases Experiment). 2011. Data provided to ERG (an EPA contractor) by Ray Wang, Georgia Institute of Technology. November 2011.

AGAGE. 2012. Monthly mean N_2O concentrations for Cape Grim, Australia. Accessed May 10, 2012. http://ds.data.jma.go.jp/gmd/wdcgg/.

Arnold, T., J. Mühle, P.K. Salameh, C.M. Harth, D.J. Ivy, and R.F. Weiss. 2012. Automated measurement of nitrogen trifluoride in ambient air. Analytical Chemistry 84(11):4798–4804.

Barnola, J.M., D. Raynaud, C. Lorius, and N.I. Barkov. 2003. Historical CO_2 record from the Vostok ice core. In Trends: A compendium of data on global change. Oak Ridge, TN: U.S. Department of Energy. Accessed September 14, 2005. http://cdiac.ornl.gov/trends/co2/vostok.html.

Battle, M., M. Bender, T. Sowers, P. Tans, J. Butler, J. Elkins, J. Ellis, T. Conway, N. Zhang, P. Lang, and A. Clarke. 1996. Atmospheric gas concentrations over the past century measured in air from firn at the South Pole. Nature 383:231–235. Accessed September 8, 2005. ftp://daac.ornl.gov/data/global_climate/global_N_cycle/data/global_N_perturbations.txt.

Blunier, T., and E.J. Brook. 2001. Timing of millennial-scale climate change in Antarctica and Greenland during the last glacial period. Science 291:109–112. Accessed September 13, 2005. ftp://ftp.ncdc.noaa.gov/pub/data/paleo/icecore/greenland/summit/grip/synchronization/readme_blunier2001.txt.

Chamard, P., L. Ciattaglia, A. di Sarra, and F. Monteleone. 2001. Atmospheric CO_2 record from flask measurements at Lampedusa Island. In Trends: A compendium of data on global change. Oak Ridge, TN: U.S. Department of Energy. Accessed September 14, 2005. http://cdiac.ornl.gov/trends/co2/lampis.html.

Dai, A., J. Wang, P.W. Thorne, D.E. Parker, L. Haimberger, and X.L. Wang. 2011. A new approach to homogenize daily radiosonde humidity data. J. Climate 24(4):965–991. http://journals.ametsoc.org/doi/abs/10.1175/2010JCLI3816.1.

Etheridge, D.M., L.P. Steele, R.J. Francey, and R.L. Langenfelds. 2002. Historical CH_4 records since about 1000 A.D. from ice core data. In Trends: A compendium of data on global change. Oak Ridge, TN: U.S. Department of Energy. Accessed September 13, 2005. http://cdiac.ornl.gov/trends/atm_meth/lawdome_meth.html.

Etheridge, D.M., L.P. Steele, R.L. Langenfelds, R.J. Francey, J.M. Barnola, and V.I. Morgan. 1998. Historical CO_2 records from the Law Dome DE08, DE08-2, and DSS ice cores. In Trends: A compendium of data on global change. Oak Ridge, TN: U.S. Department of Energy. Accessed September 14, 2005. http://cdiac.ornl.gov/trends/co2/lawdome.html.

Flückiger, J., E. Monnin, B. Stauffer, J. Schwander, T.F. Stocker, J. Chappellaz, D. Raynaud, and J.M. Barnola. 2002. High resolution Holocene N_2O ice core record and its relationship with CH_4 and CO_2. Global Biogeochem. Cycles 16(1):10–11. Accessed September 14, 2005 (N_2O), April 24, 2007 (CH_4), and April 30, 2007 (CO_2). ftp://ftp.ncdc.noaa.gov/pub/data/paleo/icecore/antarctica/epica_domec/readme_flueckiger2002.txt.

IPCC (Intergovernmental Panel on Climate Change). 2007. Climate change 2007: The physical science basis (fourth assessment report). Cambridge, UK: Cambridge University Press. www.ipcc.ch/publications_and_data/publications_ipcc_fourth_assessment_report_wg1_report_the_physical_science_basis.htm.

Machida, T., T. Nakazawa, Y. Fujii, S. Aoki, and O. Watanabe. 1995. Increase in the atmospheric nitrous oxide concentration during the last 250 years. Geophys. Res. Lett. 22(21):2921–2924. Accessed September 8, 2005. ftp://daac.ornl.gov/data/global_climate/global_N_cycle/data/global_N_perturbations.txt.

Neftel, A., H. Friedli, E. Moor, H. Lötscher, H. Oeschger, U. Siegenthaler, and B. Stauffer. 1994. Historical CO_2 record from the Siple Station ice core. In Trends: A compendium of data on global change. Oak Ridge, TN: U.S. Department of Energy. Accessed September 14, 2005. http://cdiac.ornl.gov/trends/co2/siple.html.

NOAA (National Oceanic and Atmospheric Administration). 2011a. Data provided to ERG (an EPA contractor) by Stephen Montzka, NOAA. October 2011.

NOAA. 2011b. Monthly mean CH_4 concentrations for Cape Grim, Australia. Accessed October 27, 2011. ftp://ftp.cmdl.noaa.gov/ccg/ch4/flask/month/ch4_cgo_surface-flask_1_ccgg_month.txt.

NOAA. 2012a. Annual mean CO_2 concentrations for Mauna Loa, Hawaii. Accessed May 10, 2012. ftp://ftp.cmdl.noaa.gov/ccg/co2/trends/co2_annmean_mlo.txt.

NOAA. 2012b. Monthly mean CH_4 concentrations for Mauna Loa, Hawaii. Accessed May 10, 2012. ftp://ftp.cmdl.noaa.gov/ccg/ch4/in-situ/mlo/ch4_mlo_surface-insitu_1_ccgg_month.txt.

NOAA. 2012c. Monthly mean CO_2 concentrations for Barrow, Alaska; Cape Matatula, American Samoa; and the South Pole. Accessed May 10, 2012. ftp://ftp.cmdl.noaa.gov/ccg/co2/in-situ/.

NOAA. 2012d. Monthly mean N_2O concentrations for Barrow, Alaska; Mauna Loa, Hawaii; and the South Pole. Accessed May 10, 2012. www.esrl.noaa.gov/gmd/hats/insitu/cats/cats_conc.html.

Petit, J.R., J. Jouzel, D. Raynaud, N.I. Barkov, J.M. Barnola, I. Basile, M. Bender, J. Chappellaz, M. Davis, G. Delaygue, M. Delmotte, V.M. Kotlyakov, M. Legrand, V. Lipenkov, C. Lorius, L. Pépin, C. Ritz, E. Saltzman, and M. Stievenard. 1999. Climate and atmospheric history of the past 420,000 years from the Vostok ice core, Antarctica. Nature 399:429–436. Accessed April 24, 2007. ftp://ftp.ncdc.noaa.gov/pub/data/paleo/icecore/antarctica/vostok/ch4nat.txt.

Siegenthaler, U., T. F. Stocker, E. Monnin, D. Lüthi, J. Schwander, B. Stauffer, D. Raynaud, J.M. Barnola, H. Fischer, V. Masson-Delmotte, and J. Jouzel. 2005. Stable carbon cycle-climate relationship during the late Pleistocene. Science 310(5752):1313–1317. Accessed May 15, 2007. ftp://ftp.ncdc.noaa.gov/pub/data/paleo/icecore/antarctica/epica_domec/edc-co2-650k-390k.txt.

Sowers, T., R.B. Alley, and J. Jubenville. 2003. Ice core records of atmospheric N_2O covering the last 106,000 years. Science 301(5635):945–948. Accessed September 14, 2005. www1.ncdc.noaa.gov/pub/data/paleo/icecore/antarctica/taylor/taylor_n2o.txt.

Spahni, R., J. Chappellaz, T.F. Stocker, L. Loulergue, G. Hausammann, K. Kawamura, J. Flückiger, J. Schwander, D. Raynaud, V. Masson-Delmotte, and J. Jouzel. 2005. Atmospheric methane and nitrous oxide of the late Pleistocene from Antarctic ice cores. Science 310(5752):1317–1321. Accessed May 15, 2007. ftp://ftp.ncdc.noaa.gov/pub/data/paleo/icecore/antarctica/epica_domec/edc-ch4-2005-650k.txt.

Steele, L.P., P.B. Krummel, and R.L. Langenfelds. 2002. Atmospheric CH_4 concentrations from sites in the CSIRO Atmospheric Research GASLAB air sampling network (October 2002 version). In Trends: A compendium of data on global change. Oak Ridge, TN: U.S. Department of Energy. Accessed September 13, 2005. http://cdiac.esd.ornl.gov/trends/atm_meth/csiro/csiro-shetlandch4.html.

Steele, L.P., P.B. Krummel, and R.L. Langenfelds. 2007. Atmospheric CO_2 concentrations (ppmv) derived from flask air samples collected at Cape Grim, Australia, and Shetland Islands, Scotland. Aspendale, Victoria, Australia: Atmospheric, Research, Commonwealth Scientific, and Industrial Research Organisation. Accessed January 20, 2009. http://cdiac.esd.ornl.gov/ftp/trends/co2/csiro/.

WDCGG (World Data Centre for Greenhouse Gases). 2005. Atmospheric CH_4 concentrations for Greenland Site J. Accessed September 14, 2005. http://ds.data.jma.go.jp/gmd/wdcgg/.

Weiss, R.F., J. Mühle, P.K. Salameh, and C.M. Harth. 2008. Nitrogen trifluoride in the global atmosphere. Geophys. Res. Lett. 35:L20821.

Climate Forcing

Identification

1. Indicator Description

This indicator measures the levels of greenhouse gases (GHGs) in the atmosphere between 1979 and 2010 based on their ability to cause changes in the Earth's climate. Results are reflected in the Annual Greenhouse Gas Index.

2. Revision History

April 2010: Indicator posted
December 2011: Indicator updated with data through 2010
October 2012: Indicator updated with data through 2011

Data Sources

3. Data Sources

GHG concentrations are measured by a cooperative global network of monitoring stations overseen by the National Oceanic and Atmospheric Administration's (NOAA's) Earth System Research Laboratory (ESRL). The indicator uses measurements of 20 GHGs.

4. Data Availability

This indicator is based on NOAA's Annual Greenhouse Gas Index (AGGI). Annual values of the AGGI (total and broken down by gas) are posted online at: www.esrl.noaa.gov/gmd/aggi/, along with definitions and descriptions of the data. EPA obtained data from NOAA's public website.

The AGGI is based on data from monitoring stations around the world. Most of these data were collected as part of the NOAA/ESRL cooperative monitoring network. Data files from these cooperative stations are available online at: www.esrl.noaa.gov/gmd/dv/ftpdata.html. Users can obtain station metadata by navigating to: www.esrl.noaa.gov/gmd/dv/site/, viewing a list of stations, and then selecting a station of interest.

Methane data prior to 1983 are annual averages from Etheridge et al. (1998). Users can download data from this study at: http://cdiac.ornl.gov/trends/atm_meth/lawdome_meth.html.

Methodology

5. Data Collection

This indicator is based on measurements of the concentrations of various long-lived GHGs in ambient air. These measurements have been collected following consistent high-precision techniques that have been documented in peer-reviewed literature.

The indicator uses measurements of five "major" GHGs and 15 other GHGs. The five major GHGs for this indicator are carbon dioxide (CO_2), methane (CH_4), nitrous oxide (N_2O), and two chlorofluorocarbons: CFC-11 and CFC-12. According to NOAA, these five GHGs account for about 96 percent of the increase in direct radiative forcing by long-lived GHGs since 1750. The other 15 gases are CFC-113, carbon tetrachloride (CCl_4), methyl chloroform (CH_3CCl_3), HCFC-22, HCFC-141b, HCFC-142b, HFC-23, HFC-125, HFC-134a, HFC-143a, HFC-152a, sulfur hexafluoride (SF_6), halon-1211, halon-1301, and halon-2402.

Monitoring stations in NOAA's ESRL network collect air samples at about 80 global clean air sites, although not all sites monitor for all the gases of interest. Monitoring sites include fixed stations on land as well as measurements at 5-degree latitude intervals along specific ship routes in the oceans. Monitoring stations collect data at least weekly. These weekly measurements can be averaged to arrive at an accurate representation of annual concentrations.

For a map of monitoring sites in the NOAA/ESRL cooperative network, see: www.esrl.noaa.gov/gmd/aggi. For more information about the global monitoring network and a link to an interactive map, see NOAA's website at: www.esrl.noaa.gov/gmd/dv/site.

6. Indicator Derivation

From weekly station measurements, NOAA calculated a global average concentration of each gas using a smoothed north-south latitude profile in sine latitude space. NOAA averaged these weekly global values over the course of the year to determine an annual average concentration of each gas. Pre-1983 methane measurements came from stations outside the NOAA/ESRL network; these data were adjusted to NOAA's calibration scale before being incorporated into the indicator.

Next, NOAA transformed gas concentrations into an index based on radiative forcing. These calculations account for the fact that different gases have different abilities to alter the Earth's energy balance. NOAA determined the total radiative forcing of the GHGs by applying radiative forcing factors that have been scientifically established for each gas based on its global warming potential and its atmospheric lifetime. These values and equations have been recommended by the Intergovernmental Panel on Climate Change (IPCC) (2001). In order to keep the index as accurate as possible, NOAA's radiative forcing calculations considered only direct forcing, not additional model-dependent feedbacks such as those due to water vapor and ozone depletion.

NOAA compared present-day concentrations with concentrations circa 1750 (i.e., before the start of the Industrial Revolution), and this indicator shows only the radiative forcing associated with the *increase* in concentrations since 1750. In this regard, the indicator focuses only on the additional radiative forcing that has resulted from human-influenced emissions of GHGs.

Figure 1 shows radiative forcing from the selected GHGs in units of watts per square meter. This forcing value is calculated at the tropopause, which is the boundary between the troposphere and the stratosphere. Thus, the square meter term refers to the surface area of the sphere that contains the Earth and its lower atmosphere (the troposphere). The watts term refers to the rate of energy transfer.

The data provided to EPA by NOAA also describe radiative forcing in terms of the AGGI. This unitless index is formally defined as the ratio of radiative forcing in a given year compared with a base year of 1990, which was chosen because 1990 is the baseline year for the Kyoto Protocol. Thus, 1990 is set to a total AGGI value of 1. An AGGI scale appears on the right side of Figure 1.

NOAA's monitoring network did not provide sufficient data prior to 1979, and no attempt has been made to project the indicator backward before that start date. No attempt has been made to project trends forward into the future, either.

This indicator can be reconstructed from publicly available information. NOAA's website (www.esrl.noaa.gov/gmd/aggi) provides a complete explanation of how to construct the AGGI from the available concentration data, including references to the equations used to determine each gas's contribution to radiative forcing. See Hofmann et al. (2006a) and Hofmann et al. (2006b) for more information about the AGGI and how it was constructed. See Dlugokencky et al. (2005) for information on special steps that were taken to adjust pre-1983 methane data to NOAA's calibration scale.

7. Quality Assurance and Quality Control

The online documentation for the AGGI does not explicitly discuss quality assurance and quality control procedures. NOAA's analysis has been peer-reviewed and published in the scientific literature, however (see Hofmann et al., 2006a and 2006b), and users should have confidence in the quality of the data.

Analysis

8. Comparability Over Time and Space

With the exception of pre-1983 methane measurements, all data were collected through the NOAA/ESRL global monitoring network with consistent techniques over time and space. Pre-1983 methane measurements came from stations outside the NOAA/ESRL network; these data were adjusted to NOAA's calibration scale before being incorporated into the indicator.

The data for this indicator have been spatially averaged to ensure that the final value for each year accounts for all of the original measurements to the appropriate degree. Results are considered to be globally representative, which is an appropriate assumption because the gases covered by this indicator have long residence times in the atmosphere and are considered to be well-mixed. Although there are minor variations among sampling locations, the overwhelming consistency among sampling locations indicates that extrapolation from these locations to the global atmosphere is reliable.

9. Sources of Uncertainty

This indicator is based on direct measurements of atmospheric concentrations of GHGs. These measurements are of a known and high quality, collected by a well-established monitoring network.

NOAA's AGGI website does not present explicit uncertainty values for either the AGGI or the underlying data, but exact uncertainty estimates can be obtained by contacting NOAA.

The empirical expressions used for radiative forcing are derived from atmospheric radiative transfer models and generally have an uncertainty of about 10 percent. The uncertainties in the global average concentrations of the long-lived GHGs are much smaller, according to the AGGI website documentation at: www.esrl.noaa.gov/gmd/aggi.

Uncertainty is expected to have little bearing on the conclusions for several reasons. First, the indicator is based entirely on measurements that have low inherent uncertainty. Second, the increase in GHG radiative forcing over recent years is far greater than the estimated uncertainty of underlying measurement methodologies, and it is also greater than the estimated 10 percent uncertainty in the radiative forcing equations. Thus, it is highly unlikely that the trends depicted in this indicator are somehow an artifact of uncertainties in the sampling and analytical methods.

10. Sources of Variability

Collecting data from different locations could lead to some variability, but this variability is expected to have little bearing on the conclusions. Scientists have found general agreement in trends among multiple data sets collected at different locations using different program designs, providing some assurance that the trends depicted actually represent atmospheric conditions, rather than some artifact of sampling design.

11. Statistical/Trend Analysis

The increase in GHG radiative forcing over recent years is far greater than the estimated uncertainty of underlying measurement methodologies, and it is also greater than the estimated 10 percent uncertainty in the radiative forcing equations. Thus, it is highly likely that the trends depicted in this indicator accurately represent changes in the Earth's atmosphere.

12. Data Limitations

Factors that may impact the confidence, application, or conclusions drawn from this indicator are as follows:

1. The AGGI and its underlying analysis do not provide a complete picture of radiative forcing from the major GHGs because they do not consider indirect forcing due to water vapor, ozone depletion, and other factors. These mechanisms have been excluded because quantifying them would require models that would add significant uncertainty to the indicator.
2. This indicator does not include radiative forcing due to shorter-lived GHGs and other radiatively important atmospheric constituents such as black carbon, aerosols, and sulfates. Reflective aerosol particles in the atmosphere can reduce climate forcing, for example, while tropospheric ozone can increase it. These spatially heterogeneous, short-lived climate forcing agents have uncertain global magnitudes and thus are not included in NOAA's index to maintain accuracy.

References

Dlugokencky, E.J., R.C. Myers, P.M. Lang, K.A. Masarie, A.M. Crotwell, K.W. Thoning, B.D. Hall, J.W. Elkins, and L.P Steele. 2005. Conversion of NOAA atmospheric dry air CH_4 mole fractions to a gravimetrically-prepared standard scale. J. Geophys. Res. 110:D18306.

Etheridge, D.M., L.P. Steele, R.J. Francey, and R.L. Langenfelds. 1998. Atmospheric methane between 1000 A.D. and present: Evidence of anthropogenic emissions and climate variability. J. Geophys. Res 103:15,979–15,993.

Hofmann, D.J., J.H. Butler, E.J. Dlugokencky, J.W. Elkins, K. Masarie, S.A. Montzka, and P. Tans. 2006a. The role of carbon dioxide in climate forcing from 1979–2004: Introduction of the Annual Greenhouse Gas Index. Tellus B 58B:614–619.

Hofmann, D.J., J.H. Butler, T.J. Conway, E.J. Dlugokencky, J.W. Elkins, K. Masarie, S.A. Montzka, R.C. Schnell, and P. Tans. 2006b. Tracking climate forcing: The Annual Greenhouse Gas Index. EOS, Trans. Amer. Geophys. Union 87:509–511.

IPCC (Intergovernmental Panel on Climate Change). 2001. Climate change 2001: The scientific basis (third assessment report). Cambridge, UK: Cambridge University Press. www.ipcc.ch/ipccreports/tar/wg1/index.htm.

U.S. and Global Temperature

Identification

1. Indicator Description

This indicator describes changes in average air temperature for the United States and the world from 1901 to 2011. In this indicator, temperature data are presented as trends in anomalies. Air temperature is an important component of climate, and changes in temperature can have wide-ranging direct and indirect effects on the environment and society.

Components of this indicator include:

- Changes in temperature in the contiguous 48 states over time (Figure 1)
- Changes in temperature worldwide over time (Figure 2)
- A map showing rates of temperature change across the United States (Figure 3)

2. Revision History

April 2010: Indicator posted
December 2011: Updated with data through 2010
May 2012: Updated with data through 2011

Data Sources

3. Data Sources

This indicator is based on temperature anomaly data provided by the National Oceanic and Atmospheric Administration's (NOAA's) National Climatic Data Center (NCDC).

4. Data Availability

The long-term surface time series in Figures 1, 2, and 3 were provided to EPA by NOAA's NCDC. NCDC calculated these time series based on monthly values from a set of NCDC-maintained databases: the U.S. Historical Climatology Network (USHCN) Version 2, the Global Historical Climatology Network–Monthly (GHCN-M) Version 3.1 (for global time series), and GHCN-Daily Version 2.92 (for Alaska and Hawaii maps). These databases can be accessed online. To supplement Figures 1 and 2, EPA obtained satellite-based measurements from NCDC's public website.

Contiguous 48 States (Surface)

Underlying temperature data for the contiguous 48 states come from the USHCN. Currently, the data are distributed by NCDC on various computer media (e.g., anonymous FTP sites), with no confidentiality issues limiting accessibility. Users can link to the data online at: www.ncdc.noaa.gov/oa/climate/research/ushcn/#access. Appropriate metadata and "readme" files are

appended to the data. For example, see:
ftp://ftp.ncdc.noaa.gov/pub/data/ushcn/v2/monthly/readme.txt.

Alaska, Hawaii, and Global (Surface)

GHCN temperature data can be obtained from NCDC over the Web or via anonymous FTP. This indicator is specifically based on a combined global land-sea temperature data set that can be obtained from: www.ncdc.noaa.gov/ghcnm/v3.php. There are no known confidentiality issues that limit access to the data set, and the data are accompanied by metadata.

Satellite Data

EPA obtained the satellite trends from NCDC's public website at:
www.ncdc.noaa.gov/oa/climate/research/msu.html.

Methodology

5. Data Collection

This indicator is based on temperature measurements. The global portion of this indicator presents temperatures measured over land and sea, while the portion devoted to the contiguous 48 states shows temperatures measured over land only.

Surface data for this indicator were compiled from thousands of weather stations throughout the United States and worldwide using standard meteorological instruments. Data for the contiguous 48 states were compiled in the USHCN. Data for Alaska, Hawaii, and the rest of the world were taken from the GHCN. Both of these networks are overseen by NOAA and have been extensively peer reviewed. As such, they represent the most complete long-term instrumental data sets for analyzing recent climate trends. More information on these networks can be found below.

Contiguous 48 States (Surface)

USHCN Version 2 contains monthly averaged maximum, minimum, and mean temperature data from approximately 1,200 stations within the contiguous 48 states. The period of record varies for each station but generally includes most of the 20th century. One of the objectives in establishing the USHCN was to detect secular changes of regional rather than local climate. Therefore, stations included in the network are only those believed to not be influenced to any substantial degree by artificial changes of local environments. Some of the stations in the USHCN are first-order weather stations, but the majority are selected from U.S. cooperative weather stations (approximately 5,000 in the United States). To be included in the USHCN, a station had to meet certain criteria for record longevity, data availability (percentage of available values), spatial coverage, and consistency of location (i.e., experiencing few station changes). An additional criterion, which sometimes compromised the preceding criteria, was the desire to have a uniform distribution of stations across the United States. Included with the data set are metadata files that contain information about station moves, instrumentation, observing times, and elevation. NOAA's website (www.ncdc.noaa.gov/oa/climate/research/ushcn) provides more information about USHCN data collection.

Alaska, Hawaii, and Global (Surface)

GHCN-M Version 3.1 contains monthly climate data from weather stations worldwide. Monthly mean temperature data are available for 7,280 stations, with homogeneity-adjusted data available for a subset (5,206 mean temperature stations). Data were obtained from many types of stations. For the global component of this indicator, the GHCN land-based data were merged with an additional set of long-term sea surface temperature data; this merged product is called the extended reconstructed sea surface temperature (ERSST) data set, Version #3b (Smith et al., 2008).

NCDC has published documentation for the GHCN. For more information, including data sources, methods, and recent improvements, see: www.ncdc.noaa.gov/ghcnm/v3.php and the sources listed therein. Additional background on the merged land-sea temperature data set can be found at: www.ncdc.noaa.gov/cmb-faq/anomalies.html.

Satellite Data

In Figures 1 and 2, surface measurements have been supplemented with satellite-based measurements for the period from 1979 to 2011. These satellite data were collected by NOAA's polar-orbiting satellites, which take measurements across the entire globe. Satellites equipped with the necessary measuring equipment have orbited the Earth continuously since 1978, but 1979 was the first year with complete data. This indicator uses measurements that represent the lower troposphere, which is defined here as the layer of the atmosphere extending from the Earth's surface to an altitude of about 8 kilometers.

NOAA's satellites use the Microwave Sounding Unit (MSU) to measure the intensity of microwave radiation given off by various layers of the Earth's atmosphere. The intensity of radiation is proportional to temperature, which can therefore be determined through correlations and calculations. NOAA uses different MSU channels to characterize different parts of the atmosphere. Note that since 1998, NOAA has used a newer version of the instrument called the Advanced MSU.

For more information about the methods used to collect satellite measurements, see: www.ncdc.noaa.gov/oa/climate/research/msu.html and the references cited therein.

6. Indicator Derivation

Surface Data

NOAA calculated monthly temperature means for each site. In populating the USHCN and GHCN, NOAA adjusted the data to remove biases introduced by differences in the time of observation. NOAA also employed a homogenization algorithm to identify and correct for substantial shifts in local-scale data that might reflect changes in instrumentation, station moves, or urbanization effects. These adjustments were performed according to published, peer-reviewed methods. For more information on these quality assurance and error correction procedures, see Section 7.

In this indicator, temperature data are presented as trends in anomalies. An anomaly represents the difference between an observed value and the corresponding value from a baseline period. This indicator uses a baseline period of 1901 to 2000. The choice of baseline period *will not* affect the shape or the statistical significance of the overall trend in anomalies. For temperature (absolute anomalies), it only moves the trend up or down on the graph in relation to the point defined as "zero."

To generate the temperature time series, NOAA converted measurements into monthly anomalies in degrees Fahrenheit. The monthly anomalies then were averaged to determine an annual temperature anomaly for each year.

To achieve uniform spatial coverage (i.e., not biased toward areas with a higher concentration of measuring stations), NOAA averaged anomalies within grid cells on the map to create "gridded" data sets. The graph for the contiguous 48 states (Figure 1) and the map (Figure 3) are based on an analysis using grid cells that measure 2.5 degrees latitude by 3.5 degrees longitude. The global graph (Figure 2) comes from an analysis of grid cells measuring 5 degrees by 5 degrees. These particular grid sizes have been determined to be optimal for analyzing USHCN and GHCN climate data; see: http://www.ncdc.noaa.gov/oa/climate/research/ushcn/gridbox.html for more information.

Figures 1 and 2 show trends from 1901 to 2011, based on NOAA's gridded data sets. Although earlier data are available for some stations, 1901 was selected as a consistent starting point.

The map in Figure 3 shows long-term rates of change in temperature over the United States for the period 1901–2011 except for Alaska and Hawaii, for which widespread and reliable data collection did not begin until 1918 and 1905, respectively. A regression was performed on the annual anomalies for each grid cell. Trends were calculated only in those grid cells for which data were available for at least 66 percent of the years during the full period of record. The slope of each trend (rate of temperature change per year) was calculated from the annual time series by ordinary least-squares regression and then multiplied by 100 to obtain a rate per century. No attempt has been made to portray data beyond the time and space in which measurements were made.

Satellite Data

NOAA's satellites measure microwave radiation at various frequencies, which must be converted to temperature and adjusted for time-dependent biases using a set of algorithms. Various experts recommend slightly different algorithms. Accordingly, Figure 1 and Figure 2 show globally averaged trends that have been calculated by two different organizations: the Global Hydrology and Climate Center at the University of Alabama in Huntsville (UAH) and Remote Sensing Systems (RSS). For more information about the methods used to convert satellite measurements to temperature readings for various layers of the atmosphere, see: www.ncdc.noaa.gov/oa/climate/research/msu.html and the references cited therein. Both the UAH and RSS data sets are based on updated versions of analyses that have been published in the scientific literature. For example, see Christy et al. (2000, 2003), Mears et al. (2003), and Schabel et al. (2002).

NOAA provided data in the form of monthly anomalies. EPA calculated annual anomalies, then shifted the entire curves vertically in order to display the anomalies side-by-side with surface anomalies. Shifting the curves vertically does not change the shape or magnitude of the trends; it simply results in a new baseline. No attempt has been made to portray satellite-based data beyond the time and space in which measurements were made. The satellite data in Figure 1 are restricted to the atmosphere above the contiguous 48 states.

7. Quality Assurance and Quality Control

Both the USHCN and the GHCN have undergone extensive quality assurance procedures to identify errors and biases in the data and either remove these stations from the time series or apply correction factors.

Contiguous 48 States (Surface)

Quality control procedures for the USHCN are summarized at:
www.ncdc.noaa.gov/oa/climate/research/ushcn/#processing. Homogeneity testing and data correction methods are described in numerous peer-reviewed scientific papers by NOAA's NCDC. A series of data corrections was developed to specifically address potential problems in trend estimation of the rates of warming or cooling in USHCN Version 2. They include:

- Removal of duplicate records
- Procedures to deal with missing data
- Adjusting for changes in observing practices, such as changes in observation time
- Testing and correcting for artificial discontinuities in a local station record, which might reflect station relocation, instrumentation changes, or urbanization (e.g., heat island effects)

Alaska, Hawaii, and Global (Surface)

QA/QC procedures for GHCN temperature data are described in detail in Peterson et al. (1998) and Menne and Williams (2009), and at: www.ncdc.noaa.gov/ghcnm/v3.php. GHCN data undergo rigorous QA reviews, which include pre-processing checks on source data; removal of duplicates, isolated values, and suspicious streaks; time series checks that identify spurious changes in the mean and variance via pairwise comparisons; spatial comparisons that verify the accuracy of the climatological mean and the seasonal cycle; and neighbor checks that identify outliers from both a serial and a spatial perspective.

Satellite Data

NOAA follows documented procedures for QA/QC of data from the MSU satellite instruments. For example, see NOAA's discussion of MSU calibration at:
www.star.nesdis.noaa.gov/smcd/spb/calibration/msu/msucal.pdf and:
www.star.nesdis.noaa.gov/star/documents/meetings/NIST2008/Zou_MSU_Calibration_20080114.pdf.

Analysis

8. Comparability Over Time and Space

Both the USHCN and the GHCN have undergone extensive testing to identify errors and biases in the data and either remove these stations from the time series or apply scientifically appropriate correction factors to improve the utility of the data. In particular, these corrections address changes in the time-of-day of observation, advances in instrumentation, and station location changes.

Contiguous 48 States (Surface)

Homogeneity testing and data correction methods are described in more than a dozen peer-reviewed scientific papers by NCDC. Data corrections were developed to specifically address potential problems in trend estimation of the rates of warming or cooling in the USHCN (see Section 7 for documentation). Balling and Idso (2002) compare the USHCN data with several surface and upper-air data sets and show that the effects of the various USHCN adjustments produce a significantly more positive, and likely spurious, trend in the USHCN data. In contrast, a subsequent analysis by Vose et al. (2003) found that USHCN station history information is reasonably complete and that the bias adjustment models have low residual errors.

Further analysis by Menne et al. (2009) suggests that:

> ...the collective impact of changes in observation practice at USHCN stations is systematic and of the same order of magnitude as the background climate signal. For this reason, bias adjustments are essential to reducing the uncertainty in U.S. climate trends. The largest biases in the HCN are shown to be associated with changes to the time of observation and with the widespread changeover from liquid-in-glass thermometers to the maximum minimum temperature sensor (MMTS). With respect to [USHCN] Version 1, Version 2 trends in maximum temperatures are similar while minimum temperature trends are somewhat smaller because of an apparent overcorrection in Version 1 for the MMTS instrument change, and because of the systematic impact of undocumented station changes, which were not addressed [in] Version 1.

USHCN Version 2 represents an improvement in this regard.

Some observers have expressed concerns about other aspects of station location and technology. For example, Watts (2009) expresses concern that many U.S. weather stations are sited near artificial heat sources such as buildings and paved areas, potentially biasing temperature trends over time. In response to these concerns, NOAA analyzed trends for a subset of stations that Watts had determined to be "good or best," and found the temperature trend over time to be very similar to the trend across the full set of USHCN stations (www.ncdc.noaa.gov/oa/about/response-v2.pdf). While it is true that many stations are not optimally located, NOAA's findings support the results of an earlier analysis by Peterson (2006) that found no significant bias in long-term trends associated with station siting once NOAA's homogeneity adjustments have been applied.

Alaska, Hawaii, and Global (Surface)

The GHCN applied similarly stringent criteria for data homogeneity (like the USHCN) in order to reduce bias. In acquiring data sets, the original observations were sought, and in many cases where bias was identified, the stations in question were removed from the data set. See Section 7 for documentation.

For data collected over the ocean, continuous improvement and greater spatial resolution can be expected in the coming years, with corresponding updates to the historical data. For example, there is a known bias during the World War II years (1941–1945), when almost all ocean temperature measurements were collected by U.S. Navy ships that recorded ocean intake temperatures, which can give warmer numbers than the techniques used in other years. Future efforts will aim to adjust the data more fully to account for this bias.

Satellite Data

NOAA's satellites cover the entire Earth with consistent measurement methods. Procedures to calibrate the results and correct for any biases over time are described in the references cited under Section 7.

9. Sources of Uncertainty

Surface Data

Uncertainties in temperature data increase as one goes back in time, as there are fewer stations early in the record. However, these uncertainties are not sufficient to undermine the fundamental trends in the data.

Error estimates are not readily available for U.S. temperature, but they are available for the global temperature time series. See the error bars in NOAA's graphic online at: http://www.ncdc.noaa.gov/sotc/service/global/global-land-ocean-mntp-anom/201001-201012.gif. In general, Vose and Menne (2004) suggest that the station density in the U.S. climate network is sufficient to produce a robust spatial average.

Satellite Data

Methods of inferring tropospheric temperature from satellite data have been developed and refined over time. Several independent analyses have produced largely similar curves, suggesting fairly strong agreement and confidence in the results.

Error estimates for the UAH analysis have previously been published in Christy et al. (2000, 2003). Error estimates for the RSS analysis have previously been published in Schabel et al. (2002) and Mears et al. (2003). However, error estimates are not readily available for the updated version of each analysis that EPA obtained in 2012.

10. Sources of Variability

Annual temperature anomalies naturally vary from location to location and from year to year as a result of normal variation in weather patterns, multi-year climate cycles such as the El Niño–Southern Oscillation and Pacific Decadal Oscillation, and other factors. This indicator accounts for these factors by presenting a long-term record (more than a century of data) and averaging consistently over time and space.

11. Statistical/Trend Analysis

This indicator uses ordinary least-squares regression to calculate the slope of the observed trends in temperature, but does not indicate whether each trend is statistically significant. A simple t-test indicates that some of the observed trends are significant to a 95 percent confidence level, while others are not. To conduct a more complete analysis, however, would potentially require consideration of serial correlation and other more complex statistical factors.

12. Data Limitations

Factors that may impact the confidence, application, or conclusions drawn from this indicator are as follows:

1. Biases in surface measurements may have occurred as a result of changes over time in instrumentation, measuring procedures (e.g., time of day), and the exposure and location of the instruments. Where possible, data have been adjusted to account for changes in these variables. For more information on these corrections, see Section 8. Some scientists believe that the empirical debiasing models used to adjust the data might themselves introduce non-climatic biases (e.g., Pielke et al., 2007).
2. Uncertainties in surface temperature data increase as one goes back in time, as there are fewer stations early in the record. However, these uncertainties are not sufficient to mislead the user about fundamental trends in the data.

References

Balling, Jr., R.C., and C.D. Idso. 2002. Analysis of adjustments to the United States Historical Climatology Network (USHCN) temperature database. Geophys. Res. Lett. 29(10):1387.

Christy, J.R., R.W. Spencer, and W.D. Braswell. 2000. MSU tropospheric temperatures: Dataset construction and radiosonde comparisons. J. Atmos. Oceanic Technol. 17:1153–1170. www.ncdc.noaa.gov/oa/climate/research/uah-msu.pdf.

Christy, J.R., R.W. Spencer, W.B. Norris, W.D. Braswell, and D.E. Parker. 2003. Error estimates of version 5.0 of MSU/AMSU bulk atmospheric temperatures. J. Atmos. Oceanic Technol. 20:613–629.

Mears, C.A., M.C. Schabel, and F.J. Wentz. 2003. A reanalysis of the MSU channel 2 tropospheric temperature record. J. Climate 16:3650–3664. www.ncdc.noaa.gov/oa/climate/research/rss-msu.pdf.

Menne, M.J., and C.N. Williams, Jr. 2009. Homogenization of temperature series via pairwise comparisons. J. Climate 22(7):1700–1717.

Menne, M.J., C.N. Williams, Jr., and R.S. Vose. 2009. The U.S. Historical Climatology Network monthly temperature data, version 2. Bull. Am. Meteorol. Soc. 90:993-1107. ftp://ftp.ncdc.noaa.gov/pub/data/ushcn/v2/monthly/menne-etal2009.pdf.

Peterson, T.C. 2006. Examination of potential biases in air temperature caused by poor station locations. Bull. Am. Meteorol. Soc. 87:1073–1080. http://journals.ametsoc.org/doi/pdf/10.1175/BAMS-87-8-1073.

Peterson, T.C., R. Vose, R. Schmoyer, and V. Razuvaev. 1998. Global Historical Climatology Network (GHCN) quality control of monthly temperature data. Int. J. Climatol. 18(11):1169–1179.

Pielke, R., J. Nielsen-Gammon, C. Davey, J. Angel, O. Bliss, N. Doesken, M. Cai, S. Fall, D. Niyogi, K. Gallo, R. Hale, K.G. Hubbard, X. Lin, H. Li, and S. Raman. 2007. Documentation of uncertainties and biases associated with surface temperature measurement sites for climate change assessment. Bull. Am. Meteorol. Soc. 88:913–928.

Schabel, M.C., C.A. Mears, and F.J. Wentz. 2002. Stable long-term retrieval of tropospheric temperature time series from the Microwave Sounding Unit. Proceedings of the International Geophysics and Remote Sensing Symposium III:1845–1847.

Smith, T.M., R.W. Reynolds, T.C. Peterson, and J. Lawrimore. 2008. Improvements to NOAA's historical merged land–ocean surface temperature analysis (1880–2006). J. Climate 21:2283–2296. www.ncdc.noaa.gov/ersst/papers/SEA.temps08.pdf.

Vose, R.S., and M.J. Menne. 2004. A method to determine station density requirements for climate observing networks. J. Climate 17(15):2961–2971.

Vose, R.S., C.N. Williams, Jr., T.C. Peterson, T.R. Karl, and D.R. Easterling. 2003. An evaluation of the time of observation bias adjustment in the U.S. Historical Climatology Network. Geophys. Res. Lett. 30(20):2046.

Watts, A. 2009. Is the U.S. surface temperature record reliable? The Heartland Institute. http://wattsupwiththat.files.wordpress.com/2009/05/surfacestationsreport_spring09.pdf.

High and Low Temperatures

Identification

1. Indicator Description

This indicator describes trends in unusually hot and cold temperatures across the United States over approximately the last 100 years. Extreme temperature events like summer heat waves and winter cold spells can have profound effects on society.

Components of this indicator include:

- An index reflecting the frequency of extreme heat events (Figure 1)
- The percentage of land area experiencing unusually hot summer temperatures or unusually cold winter temperatures (Figures 2 and 3, respectively)
- The proportion of record-setting high temperatures to record low temperatures over time (Figure 4)

2. Revision History

April 2010: Indicator posted
December 2011: Updated Figure 1 with data through 2010; combined Figures 2 and 3 into a new Figure 2, and updated data through 2011; added new Figures 3 and 4; and expanded the indicator from "Heat Waves" to "High and Low Temperatures"
February 2012: Updated Figure 1 with data through 2011
March 2012: Updated Figure 3 with data through 2012
October 2012: Updated Figure 2 with data through 2012

Data Sources

3. Data Sources

Index values for Figure 1 were provided by Dr. Kenneth Kunkel of the National Oceanic and Atmospheric Administration's (NOAA's) Cooperative Institute for Climate and Satellites (CICS), who updated an analysis that was previously published in U.S. Climate Change Science Program (2008). Data for Figures 2 and 3 come from NOAA's U.S. Climate Extremes Index (CEI). Data for Figure 4 come from an analysis published by Meehl et al. (2009).

All components of this indicator are based on temperature measurements from weather stations overseen by NOAA's National Weather Service (NWS). These underlying data are maintained by NCDC.

4. Data Availability

Figure 1. U.S. Annual Heat Wave Index, 1895–2011

Data for this figure were provided by Dr. Kenneth Kunkel of NOAA CICS, who performed the analysis based on data from NCDC's publicly available databases.

Figures 2 and 3. Area of the Contiguous 48 States with Unusually Hot Summer Temperatures (1910–2012) or Unusually Cold Winter Temperatures (1911–2012)

NOAA has calculated each of the components of the CEI and has made these data files publicly available. The data for unusually hot summer maximum and minimum temperatures (CEI steps 1b and 2b) can be downloaded from: ftp://ftp.ncdc.noaa.gov/pub/data/cei/dk-step1-hi.06-08.results and: ftp://ftp.ncdc.noaa.gov/pub/data/cei/dk-step2-hi.06-08.results, respectively. The data for unusually cold winter maximum and minimum temperatures (CEI steps 1a and 2a) can be downloaded from: ftp://ftp.ncdc.noaa.gov/pub/data/cei/dk-step1-lo.12-02.results and: ftp://ftp.ncdc.noaa.gov/pub/data/cei/dk-step2-lo.12-02.results, respectively. A "readme" file (ftp://ftp.ncdc.noaa.gov/pub/data/cei) explains the contents of the data files. NOAA's CEI website (http://www.ncdc.noaa.gov/extremes/cei/) provides additional descriptions and links, along with a portal to download or graph various components of the CEI, including the data sets listed above.

Figure 4. Record Daily High and Low Temperatures in the Contiguous 48 States, 1950–2009

Ratios of record highs to lows were taken from Meehl et al. (2009) and a press release that accompanied the publication of that peer-reviewed study (http://www2.ucar.edu/news/1036/record-high-temperatures-far-outpace-record-lows-across-us). For confirmation, EPA obtained the actual counts of highs and lows by decade from Claudia Tebaldi, a co-author of the Meehl et al. (2009) paper.

Underlying Data

NCDC maintains a set of databases that provide public access to daily and monthly temperature records from thousands of weather stations across the country. For access to these data and accompanying metadata, visit NCDC's website at: http://www.ncdc.noaa.gov/oa/ncdc.html.

Many of the weather stations are part of NOAA's Cooperative Observer Program (COOP). Complete data, embedded definitions, and data descriptions for these stations can be found online at: www.ncdc.noaa.gov/doclib/. State-specific data can be found at: www7.ncdc.noaa.gov/IPS/coop/coop.html;jsessionid=312EC0892FFC2FBB78F63D0E3ACF6CBC. There are no confidentiality issues that may limit accessibility. Additional metadata can be found at: www.nws.noaa.gov/om/coop/.

Methodology

5. Data Collection

Systematic collection of weather data in the United States began in the 1800s. Since then, observations have been recorded from 23,000 stations. At any given time, observations are recorded from approximately 8,000 stations. Observations are made on an hourly basis, and the maximum and minimum temperatures are recorded for each 24-hour time span.

NOAA's National Weather Service (NWS) operates some stations (called first-order stations), but the vast majority of U.S. weather stations are part of NWS's Cooperative Observer Program (COOP). The COOP data set represents the core climate network of the United States (Kunkel et al., 2005). Cooperative observers include state universities, state and federal agencies, and private individuals. Observers are trained to collect data following NWS protocols, and equipment to gather these data is provided and maintained by the NWS.

Data collected by COOP are referred to as U.S. Daily Surface Data or Summary of the Day data. Variables that are relevant to this indicator include observations of daily maximum and minimum temperatures. General information about the NWS COOP data set is available at: www.nws.noaa.gov/os/coop/what-is-coop.html. Sampling procedures are described in Kunkel et al. (2005) and in the full metadata for the COOP data set available at: www.nws.noaa.gov/om/coop/.

NCDC also maintains a database called the U.S. Historical Climatology Network (USHCN), which contains data from a subset of COOP and first-order weather stations that meet certain selection criteria and undergo additional levels of quality control. USHCN contains monthly averaged maximum, minimum, and mean temperature data from approximately 1,200 stations within the contiguous 48 states. The period of record varies for each station but generally includes most of the 20th century. One of the objectives in establishing the USHCN was to detect secular changes of regional rather than local climate. Therefore, stations included in this network are only those believed to not be influenced to any substantial degree by artificial changes of local environments. To be included in the USHCN, a station had to meet certain criteria for record longevity, data availability (percentage of available values), spatial coverage, and consistency of location (i.e., experiencing few station changes). An additional criterion, which sometimes compromised the preceding criteria, was the desire to have a uniform distribution of stations across the United States. Included with the data set are metadata files that contain information about station moves, instrumentation, observing times, and elevation. NOAA's website (www.ncdc.noaa.gov/oa/climate/research/ushcn) provides more information about USHCN data collection.

All four figures use data from the contiguous 48 states. Original sources and selection criteria are as follows:

- Figure 1 is based on stations from the COOP data set that had sufficient data during the period of record (1895–2011).
- Figures 2 and 3 are based on the narrower set of stations contained within the USHCN, which is the source of all data for NOAA's CEI. Additional selection criteria were applied to these data prior to inclusion in CEI calculations, as described by Gleason et al. (2008). In compiling the temperature components of the CEI, NOAA selected only those stations with monthly temperature data at least 90 percent complete within a given period (e.g., annual, seasonal) as well as 90 percent complete for the full period of record.
- In Figure 4, data for the 1950s through 1990s are based on a subset of 2,000 COOP stations that have collected data since 1950 and had no more than 10 percent missing values during the period from 1950 to 2006. These selection criteria are further described in Meehl et al. (2009).
- In Figure 4, data for the 2000s are based on the complete set of COOP records available from 2000 through September 2009. These numbers were published in Meehl et al. (2009) and the

accompanying press release, but they do not follow the same selection criteria as the previous decades (as described above). Counts of record highs and lows using the Meehl et al. (2009) selection criteria were available, but only through 2006. Thus, to make this indicator as current as possible, EPA chose to use data from the broader set that extends through September 2009. Using the 2000–2006 data would result in a high:low ratio of 1.86, compared with a ratio of 2.04 when the full-decade data set (shown in Figure 4) is considered.

6. Indicator Derivation

Figure 1. U.S. Annual Heat Wave Index, 1895–2011

Data from the COOP data set have been used to calculate annual values for a U.S. Annual Heat Wave Index. In this indicator, heat waves are defined as warm periods of at least four days with an average temperature (that is, averaged over all four days) exceeding the threshold for a one-in-10-year occurrence (Kunkel et al., 1999). The Annual U.S. Heat Wave Index is a frequency measure of the number of heat waves that occur each year. A complete explanation of trend analysis in the annual average heat wave index values, especially trends occurring since 1960, can be found in Appendix A, Example 2, of U.S. Climate Change Science Program (2008). Analytical procedures are described in Kunkel et al. (1999).

Figures 2 and 3. Area of the Contiguous 48 States with Unusually Hot Summer Temperatures (1910–2012) or Unusually Cold Winter Temperatures (1911–2012)

Figure 2 of this indicator shows the percentage of the area of the contiguous 48 states in any given year that experienced unusually warm maximum and minimum summer temperatures. Figure 3 displays the percentage of land area that experienced unusually cold maximum and minimum winter temperatures.

Figures 2 and 3 were developed as subsets of NOAA's CEI, an index that uses six variables to examine trends in extreme weather and climate. These figures are based on components of NOAA's CEI (labeled as Steps 1a, 1b, 2a, and 2b) that look at the percentage of land area within the contiguous 48 states that experienced maximum (Step 1) or minimum (Step 2) temperatures much below (a) or above (b) normal.

NOAA computed the data for the CEI and calculated the percentage of land area for each year by dividing the contiguous 48 states into a 1-degree by 1-degree grid and using data from one station per grid box. This was done to eliminate many of the artificial extremes that resulted from a changing number of available stations over time.

NOAA began by averaging all daily highs at a given station over the course of a month to derive a monthly average high, then performing the same step with daily lows. Next, period (monthly) averages were sorted and ranked, and values were identified as "unusually warm" if they fell in the highest 10th percentile in the period of record for each station or grid cell, and "unusually cold" if they fell in the lowest 10th percentile. Thus, the CEI has been constructed to have an expected value of 10 percent for each of these components based on the historical record—or a value of 20 percent if the two "extreme" ends of the distribution are added together.

The CEI can be calculated for individual months, seasons, or an entire year. Figure 2 displays data for summer, which the CEI defines as June, July, and August. Figure 3 displays data for winter, which the CEI defines as December, January, and February. Winter values are plotted at the year in which the season ended; for example, the winter from December 2010 to February 2011 is plotted at year 2011. This explains why Figures 2 and 3 appear to have a different starting year, as data were not available from December 1909 to calculate a winter value for 1910. To smooth out some of the year-to-year variability, EPA applied a nine-point binomial filter, which is plotted at the center of each nine-year window. For example, the smoothed value from 2002 to 2010 is plotted at year 2006. NOAA NCDC recommends this approach and has used it in the official online reporting tool for the CEI.

EPA used endpoint padding to extend the nine-year smoothed lines all the way to the ends of the period of record. As recommended by NCDC, EPA calculated smoothed values as follows: If 2012 was the most recent year with data available, EPA calculated smoothed values to be centered at 2009, 2010, 2011, and 2012 by inserting the 2012 data point into the equation in place of the as-yet-unreported annual data points for 2013 and beyond. EPA used an equivalent approach at the beginning of the time series.

The CEI has been extensively documented and refined over time to provide the best possible representation of trends in extreme weather and climate. For an overview of how NOAA constructed Steps 1 and 2 of the CEI, see: www.ncdc.noaa.gov/oa/climate/research/cei/cei.html. This page provides a list of references that describe analytical methods in greater detail. In particular, see Gleason et al. (2008).

Figure 4. Record Daily High and Low Temperatures in the Contiguous 48 States, 1950–2009

Figure 4 displays the proportion of daily record high and daily record low temperatures reported at a subset of quality-controlled NCDC COOP network stations (except for the most recent decade, which is based on the entire COOP network as described in Section 5). As described in Meehl et al. (2009), steps were taken to fill missing data points with simple averages from neighboring days with reported values when there are no more than two consecutive days missing, or otherwise by interpolating values at the closest surrounding stations.

Based on the total number of record highs and the total number of record lows set in each decade, Meehl et al. (2009) calculated each decade's ratio of record highs to record lows. EPA converted these values to percentages to make the results easier to communicate.

Although it might be interesting to look at trends in the absolute number of record highs and record lows over time, these values are recorded in a way that would make a trend analysis misleading. A daily high or low is registered as a "record" if it broke a record *at the time*—even if that record has since been surpassed. Statistics dictate that as more years go by, it becomes less likely that a record will be broken. In contrast, if a station has only been measuring temperature for 5 years (for example), every day has a much greater chance of breaking a previous record. Thus, a decreasing trend in absolute counts does not indicate that the climate is actually becoming less extreme, as one might initially guess. Meehl et al. (2009) show that actual counts indeed fit a decreasing pattern over time, as expected statistically.

7. Quality Assurance and Quality Control

The NWS has documented COOP methods, including training manuals and maintenance of equipment, at: www.nws.noaa.gov/os/coop/training.htm. These training materials also discuss quality control of the underlying data set. Additionally, pre-1948 data in the COOP data set have recently been digitized from hard copy. Quality control procedures associated with digitization and other potential sources of error are discussed in Kunkel et al. (2005).

Quality control procedures for the USHCN are summarized at:
www.ncdc.noaa.gov/oa/climate/research/ushcn/#processing. Homogeneity testing and data correction methods are described in numerous peer-reviewed scientific papers by NCDC. A series of data corrections was developed to specifically address potential problems in trend estimation of the rates of warming or cooling in USHCN Version 2. They include:

- Removal of duplicate records
- Procedures to deal with missing data
- Adjusting for changes in observing practices, such as changes in observation time
- Testing and correcting for artificial discontinuities in a local station record, which might reflect station relocation, instrumentation changes, or urbanization (e.g., heat island effects)

Analysis

8. Comparability Over Time and Space

Long-term weather stations have been carefully selected from the full set of all COOP stations to provide an accurate representation of the United States for the U.S. Annual Heat Wave Index and the proportion of record daily highs to record daily lows (Kunkel et al., 1999; Meehl et al., 2009). Some bias may have occurred as a result of changes over time in instrumentation, measuring procedures, and the exposure and location of the instruments. The record high/low analysis begins at 1950 in an effort to reduce disparity in station record lengths.

The USHCN has undergone extensive testing to identify errors and biases in the data and either remove these stations from the time series or apply scientifically appropriate correction factors to improve the utility of the data. In particular, these corrections address changes in the time-of-day of observation, advances in instrumentation, and station location changes.

Homogeneity testing and data correction methods are described in more than a dozen peer-reviewed scientific papers by NCDC. Data corrections were developed to specifically address potential problems in trend estimation of the rates of warming or cooling in the USHCN (see Section 7 for documentation). Balling and Idso (2002) compare the USHCN data with several surface and upper-air data sets and show that the effects of the various USHCN adjustments produce a significantly more positive, and likely spurious, trend in the USHCN data. In contrast, a subsequent analysis by Vose et al. (2003) found that USHCN station history information is reasonably complete and that the bias adjustment models have low residual errors.

Further analysis by Menne et al. (2009) suggests that:

> ...the collective impact of changes in observation practice at USHCN stations is systematic and of the same order of magnitude as the background climate signal. For this reason, bias adjustments are essential to reducing the uncertainty in U.S. climate trends. The largest biases in the HCN are shown to be associated with changes to the time of observation and with the widespread changeover from liquid-in-glass thermometers to the maximum minimum temperature sensor (MMTS). With respect to

[USHCN] Version 1, Version 2 trends in maximum temperatures are similar while minimum temperature trends are somewhat smaller because of an apparent overcorrection in Version 1 for the MMTS instrument change, and because of the systematic impact of undocumented station changes, which were not addressed [in] Version 1.

USHCN Version 2 represents an improvement in this regard.

Some observers have expressed concerns about other aspects of station location and technology. For example, Watts (2009) expresses concern that many U.S. weather stations are sited near artificial heat sources such as buildings and paved areas, potentially biasing temperature trends over time. In response to these concerns, NOAA analyzed trends for a subset of stations that Watts had determined to be "good or best," and found the temperature trend over time to be very similar to the trend across the full set of USHCN stations (www.ncdc.noaa.gov/oa/about/response-v2.pdf). While it is true that many stations are not optimally located, NOAA's findings support the results of an earlier analysis by Peterson (2006) that found no significant bias in long-term trends associated with station siting once NOAA's homogeneity adjustments have been applied.

9. Sources of Uncertainty

Uncertainty may be introduced into this data set when hard copies of historical data are digitized. As a result of these and other reasons, uncertainties in the temperature data increase as one goes back in time, particularly given that there are fewer stations early in the record. However, NOAA does not believe these uncertainties are sufficient to undermine the fundamental trends in the data. Vose and Menne (2004) suggest that the station density in the U.S. climate network is sufficient to produce robust spatial averages.

Error estimates have been developed for certain segments of the data set, but do not appear to be available for the data set as a whole. Uncertainty measurements are not included with the publication of the U.S. Annual Heat Wave Index or the CEI seasonal temperature data. Error measurements for the pre-1948 COOP data set are discussed in detail in Kunkel et al. (2005).

10. Sources of Variability

Inter-annual temperature variability results from normal year-to-year variation in weather patterns, multi-year climate cycles such as the El Niño–Southern Oscillation and Pacific Decadal Oscillation, and other factors. This indicator presents nine-year smoothed curves (Figures 1, 2, and 3) and decadal averages (Figure 4) to reduce the year-to-year "noise" inherent in the data.

11. Statistical/Trend Analysis

Heat wave trends are somewhat difficult to analyze because of the presence of several outlying values in the 1930s. Statistical methods used to analyze trends in the U.S. Annual Heat Wave Index are presented in Appendix A, Example 2, of U.S. Climate Change Science Program (2008). Despite the presence of inter-annual variability and several outlying values in the 1930s, standard statistical treatments can be applied to assess a highly statistically significant linear trend from 1960 to 2011. However, the trend over the full period of record is not statistically significant.

This indicator does not report on the slope of the apparent trends in Figures 2, 3, and 4, nor does it calculate the statistical significance of these trends.

12. Data Limitations

Factors that may impact the confidence, application, or conclusions drawn from this indicator are as follows:

1. Biases may have occurred as a result of changes over time in instrumentation, measuring procedures, and the exposure and location of the instruments. Where possible, data have been adjusted to account for changes in these variables. For more information on these corrections, see Section 7.
2. Observer errors, such as errors in reading instruments or writing observations on the form, are present in the earlier part of this data set. Additionally, uncertainty may be introduced into this data set when hard copies of data are digitized. As a result of these and other reasons, uncertainties in the temperature data increase as one goes back in time, particularly given that there are fewer stations early in the record. However, NOAA does not believe these uncertainties are sufficient to undermine the fundamental trends in the data. More information about limitations of early weather data can be found in Kunkel et al. (2005).

References

Balling, Jr., R.C., and C.D. Idso. 2002. Analysis of adjustments to the United States Historical Climatology Network (USHCN) temperature database. Geophys. Res. Lett. 29(10):1387.

Gleason, K.L., J.H. Lawrimore, D.H. Levinson, T.R. Karl, and D.J. Karoly. 2008. A revised U.S. climate extremes index. J. Climate 21:2124–2137.

Kunkel, K.E., R.A. Pielke Jr., and S. A. Changnon. 1999. Temporal fluctuations in weather and climate extremes that cause economic and human health impacts: A review. Bull. Am. Meteorol. Soc. 80:1077–1098.

Kunkel, K.E., D.R. Easterling, K. Hubbard, K. Redmond, K. Andsager, M.C. Kruk, and M.L. Spinar. 2005. Quality control of pre-1948 Cooperative Observer Network data. J. Atmos. Oceanic Technol. 22:1691–1705.

Meehl, G.A., C. Tebaldi, G. Walton, D. Easterling, and L. McDaniel. 2009. Relative increase of record high maximum temperatures compared to record low minimum temperatures in the U.S. Geophys. Res. Lett. 36:L23701.

Menne, M.J., C.N. Williams, Jr., and R.S. Vose. 2009. The U.S. Historical Climatology Network monthly temperature data, version 2. Bull. Am. Meteorol. Soc. 90:993-1107. ftp://ftp.ncdc.noaa.gov/pub/data/ushcn/v2/monthly/menne-etal2009.pdf.

Peterson, T.C. 2006. Examination of potential biases in air temperature caused by poor station locations. Bull. Am. Meteorol. Soc. 87:1073–1080. http://journals.ametsoc.org/doi/pdf/10.1175/BAMS-87-8-1073.

U.S. Climate Change Science Program. 2008. Synthesis and Assessment Product 3.3: Weather and climate extremes in a changing climate. www.climatescience.gov/Library/sap/sap3-3/final-report/sap3-3-final-Chapter2.pdf.

Vose, R.S., and M.J. Menne. 2004. A method to determine station density requirements for climate observing networks. J. Climate 17(15):2961-2971.

Vose, R.S., C.N. Williams, Jr., T.C. Peterson, T.R. Karl, and D.R. Easterling. 2003. An evaluation of the time of observation bias adjustment in the U.S. Historical Climatology Network. Geophys. Res. Lett. 30(20):2046.
Watts, A. 2009. Is the U.S. surface temperature record reliable? The Heartland Institute. http://wattsupwiththat.files.wordpress.com/2009/05/surfacestationsreport_spring09.pdf.

U.S. and Global Precipitation

Identification

1. Indicator Description

This indicator describes changes in total precipitation over land for the United States and the world from 1901 to 2011. In this indicator, precipitation data are presented as trends in anomalies. Precipitation is an important component of climate, and changes in precipitation can have wide-ranging direct and indirect effects on the environment and society.

Components of this indicator include:

- Changes in precipitation in the contiguous 48 states over time (Figure 1)
- Changes in worldwide precipitation over land through time (Figure 2)
- A map showing rates of precipitation change across the United States (Figure 3)

2. Revision History

April 2010: Indicator posted
December 2011: Updated with data through 2010
May 2012: Updated with data through 2011

Data Sources

3. Data Sources

This indicator is based on precipitation anomaly data provided by the National Oceanic and Atmospheric Administration's (NOAA's) National Climatic Data Center (NCDC).

4. Data Availability

Data for this indicator were provided to EPA by NOAA's NCDC. NCDC calculated these time series based on monthly values from two NCDC-maintained databases: the U.S. Historical Climatology Network (USHCN) Version 2 and the Global Historical Climatology Network–Monthly (GHCN-M) Version 2. Both of these databases can be accessed online.

Contiguous 48 States

Underlying precipitation data for the contiguous 48 states come from the USHCN. Currently, the data are distributed by NCDC on various computer media (e.g., anonymous FTP sites), with no confidentiality issues limiting accessibility. Users can link to the data online at: www.ncdc.noaa.gov/oa/climate/research/ushcn. Appropriate metadata and "readme" files are appended to the data. For example, see: ftp://ftp.ncdc.noaa.gov/pub/data/ushcn/v2/monthly/readme.txt.

Alaska, Hawaii, and Global

GHCN precipitation data can be obtained from NCDC over the Web or via anonymous FTP. For access to GHCN data, see: www.ncdc.noaa.gov/ghcnm/v2.php. There are no known confidentiality issues that limit access to the data set, and the data are accompanied by metadata.

Methodology

5. Data Collection

This indicator is based on precipitation measurements collected from thousands of weather stations throughout the United States and over land worldwide using standard meteorological instruments. Data for the contiguous 48 states were compiled in the USHCN. Data for Alaska, Hawaii, and the rest of the world were taken from the GHCN. Both of these networks are overseen by NOAA and have been extensively peer reviewed. As such, they represent the most complete long-term instrumental data sets for analyzing recent climate trends. More information on these networks can be found below.

Contiguous 48 States

USHCN Version 2 contains total monthly precipitation data from approximately 1,200 stations within the contiguous 48 states. The period of record varies for each station but generally includes most of the 20th century. One of the objectives in establishing the USHCN was to detect secular changes of regional rather than local climate. Therefore, stations included in the network are only those believed to not be influenced to any substantial degree by artificial changes of local environments. Some of the stations in the USHCN are first-order weather stations, but the majority are selected from U.S. cooperative weather stations (approximately 5,000 in the United States). To be included in the USHCN, a station had to meet certain criteria for record longevity, data availability (percentage of available values), spatial coverage, and consistency of location (i.e., experiencing few station changes). An additional criterion, which sometimes compromised the preceding criteria, was the desire to have a uniform distribution of stations across the United States. Included with the data set are metadata files that contain information about station moves, instrumentation, observing times, and elevation. NOAA's website (www.ncdc.noaa.gov/oa/climate/research/ushcn) provides more information about USHCN data collection.

Alaska, Hawaii, and Global

GHCN Version 2 contains monthly climate data from 20,590 weather stations worldwide. Data were obtained from many types of stations.

NCDC has published documentation for the GHCN. For more information, including data sources, methods, and recent improvements, see: www.ncdc.noaa.gov/ghcnm/v2.php and the sources listed therein.

6. Indicator Derivation

NOAA calculated monthly precipitation totals for each site. In populating the USHCN and GHCN, NOAA employed a homogenization algorithm to identify and correct for substantial shifts in local-scale data that might reflect changes in instrumentation, station moves, or urbanization effects. These adjustments were performed according to published, peer-reviewed methods. For more information on these quality assurance and error correction procedures, see Section 7.

In this indicator, precipitation data are presented as trends in anomalies. An anomaly represents the difference between an observed value and the corresponding value from a baseline period. This indicator uses a baseline period of 1901 to 2000. The choice of baseline period will not affect the shape or the statistical significance of the overall trend in anomalies. For precipitation (percentage anomalies), it moves the curve up or down and may change the magnitude slightly.

To generate the precipitation time series, NOAA converted measurements into anomalies for total monthly precipitation, in millimeters. Monthly anomalies were added to find an annual anomaly for each year, which was then converted to a percent anomaly—i.e., the percent departure from the average annual precipitation during the baseline period.

To achieve uniform spatial coverage (i.e., not biased toward areas with a higher concentration of measuring stations), NOAA averaged anomalies within grid cells on the map to create "gridded" data sets. The graph for the contiguous 48 states (Figure 1) and the map (Figure 3) are based on an analysis using grid cells that measure 2.5 degrees latitude by 3.5 degrees longitude. The global graph (Figure 2) comes from an analysis of grid cells measuring 5 degrees by 5 degrees. These particular grid sizes have been determined to be optimal for analyzing USHCN and GHCN climate data; see: http://www.ncdc.noaa.gov/oa/climate/research/ushcn/gridbox.html for more information.

Figures 1 and 2 show trends from 1901 to 2011, based on NOAA's gridded data sets. Although earlier data are available for some stations, 1901 was selected as a consistent starting point.

The map in Figure 3 shows long-term rates of change in precipitation over the United States for the 1901–2011 period except for Alaska and Hawaii, for which widespread and reliable data collection did not begin until 1918 and 1905, respectively. A regression was performed on the annual anomalies for each grid cell. Trends were calculated only in those grid cells for which data were available for at least 66 percent of the years during the full period of record. The slope of each trend (percent change in precipitation per year) was calculated from the annual time series by ordinary least-squares regression and then multiplied by 100 to obtain a rate per century. No attempt has been made to portray data beyond the time and space in which measurements were made.

7. Quality Assurance and Quality Control

Both the USHCN and the GHCN have undergone extensive quality assurance procedures to identify errors and biases in the data and either remove these stations from the time series or apply correction factors.

Contiguous 48 States

Quality control procedures for the USHCN are summarized at:
www.ncdc.noaa.gov/oa/climate/research/ushcn/#processing. Homogeneity testing and data correction methods are described in numerous peer-reviewed scientific papers by NOAA's NCDC. A series of data corrections was developed to specifically address potential problems in trend estimation in USHCN Version 2. They include:

- Removal of duplicate records
- Procedures to deal with missing data
- Testing and correcting for artificial discontinuities in a local station record, which might reflect station relocation or instrumentation changes

Alaska, Hawaii, and Global

QA/QC procedures for GHCN precipitation data are described at: www.ncdc.noaa.gov/ghcnm/v2.php. GHCN data undergo rigorous quality assurance reviews, which include pre-processing checks on source data; removal of duplicates, isolated values, and suspicious streaks; time series checks that identify spurious changes in the mean and variance; spatial comparisons that verify the accuracy of the climatological mean and the seasonal cycle; and neighbor checks that identify outliers from both a serial and a spatial perspective.

Analysis

8. Comparability Over Time and Space

Both the USHCN and the GHCN have undergone extensive testing to identify errors and biases in the data and either remove these stations from the time series or apply scientifically appropriate correction factors to improve the utility of the data. In particular, these corrections address advances in instrumentation and station location changes. See Section 7 for documentation.

9. Sources of Uncertainty

Uncertainties in precipitation data increase as one goes back in time, as there are fewer stations early in the record. However, these uncertainties are not sufficient to undermine the fundamental trends in the data.

Error estimates are not readily available for U.S. or global precipitation. Vose and Menne (2004) suggest that the station density in the U.S. climate network is sufficient to produce a robust spatial average.

10. Sources of Variability

Annual precipitation anomalies naturally vary from location to location and from year to year as a result of normal variation in weather patterns, multi-year climate cycles such as the El Niño–Southern Oscillation and Pacific Decadal Oscillation, and other factors. This indicator accounts for these factors by presenting a long-term record (more than a century of data) and averaging consistently over time and space.

11. Statistical/Trend Analysis

This indicator uses ordinary least-squares regression to calculate the slope of the observed trends in precipitation, but does not indicate whether each trend is statistically significant. A simple t-test indicates that some of the observed trends are significant to a 95 percent confidence level, while others are not. To conduct a more complete analysis, however, would potentially require consideration of serial correlation and other more complex statistical factors.

12. Data Limitations

Factors that may impact the confidence, application, or conclusions drawn from this indicator are as follows:

1. Biases in measurements may have occurred as a result of changes over time in instrumentation, measuring procedures, and the exposure and location of the instruments. Where possible, data have been adjusted to account for changes in these variables. For more information on these corrections, see Section 7.
2. Uncertainties in precipitation data increase as one goes back in time, as there are fewer stations early in the record. However, these uncertainties are not sufficient to undermine the fundamental trends in the data.

References

Vose, R.S., and M.J. Menne. 2004. A method to determine station density requirements for climate observing networks. J. Climate 17(15):2961–2971.

Heavy Precipitation

Identification

1. Indicator Description

This indicator tracks the frequency of heavy precipitation events in the United States between 1895 and 2011. The potential impacts of heavy precipitation include crop damage, soil erosion, flooding, and diminished water quality.

Components of this indicator include:

- Percent of land area in the contiguous 48 states experiencing abnormal amounts of annual rainfall from one-day precipitation events (Figure 1)
- Percent of land area in the contiguous 48 states with unusually high annual precipitation (Figure 2)

2. Revision History

April 2010: Indicator posted
December 2011: Updated with data through 2010
March 2012: Updated with data through 2011

Data Sources

3. Data Sources

This indicator is based on precipitation measurements collected at weather stations throughout the contiguous 48 states. Most of the stations are part of the U.S. Historical Climatology Network (USHCN), a database compiled and managed by the National Oceanic and Atmospheric Administration's (NOAA's) National Climatic Data Center (NCDC). Indicator data were obtained from NCDC.

4. Data Availability

USHCN precipitation data are maintained at NOAA's NCDC, and the data are distributed on various computer media (e.g., anonymous FTP sites), with no confidentiality issues limiting accessibility. Users can link to the data online at: www.ncdc.noaa.gov/oa/climate/research/ushcn/#access.

Appropriate metadata and "readme" files are appended to the data so that they are discernible for analysis. For example, see: ftp://ftp.ncdc.noaa.gov/pub/data/ushcn/v2/monthly/readme.txt.

Figure 1. Extreme One-Day Precipitation Events in the Contiguous 48 States, 1910–2011

NOAA has calculated each of the components of the U.S. Climate Extremes Index (CEI) and has made these data files publicly available. The data set for extreme precipitation (CEI step 4) can be downloaded from: ftp://ftp.ncdc.noaa.gov/pub/data/cei/dk-step4.01-12.results. A "readme" file (at ftp://ftp.ncdc.noaa.gov/pub/data/cei) explains the contents of the data files.

Figure 2. Unusually High Annual Precipitation in the Contiguous 48 States, 1895–2011

Standardized Precipitation Index (SPI) data are publicly available and can be downloaded from: ftp://ftp.ncdc.noaa.gov/pub/data/cirs. This indicator uses 12-month SPI data, which are found in the file "drd964x.sp12.txt." This FTP site also includes a "readme" file that explains the contents of the data files.

Constructing Figure 2 required additional information about the U.S. climate divisions. The land area of each climate division can be found by going to: www.ncdc.noaa.gov/oa/climate/surfaceinventories.html and viewing the "U.S. climate divisions" file (exact link: ftp://ftp.ncdc.noaa.gov/pub/data/inventories/DIV-AREA.TXT). For a guide to the numerical codes assigned to each state, see: ftp://ftp.ncdc.noaa.gov/pub/data/inventories/COOP-STATE-CODES.TXT.

Methodology

5. Data Collection

This indicator is based on precipitation measurements collected by a network of thousands of weather stations spread throughout the contiguous 48 states. These stations are currently overseen by NOAA, and they use standard gauges to measure the amount of precipitation received on a daily basis. Some of the stations in the USHCN are first-order weather stations, but the majority are selected from U.S. cooperative weather stations (approximately 5,000 in the United States).

NOAA's NCDC has published extensive documentation about data collection methods for the USHCN data set. See: www.ncdc.noaa.gov/oa/climate/research/ushcn, which lists a set of technical reports and peer-reviewed articles that provide more detailed information about USHCN methodology. See: www.ncdc.noaa.gov/oa/ncdc.html for information on other types of weather stations that have been used to supplement the USHCN record.

6. Indicator Derivation

Figure 1 and Figure 2 are based on similar raw data (i.e., daily precipitation measurements), but were developed using two different models because they show trends in extreme precipitation from two different perspectives.

Figure 1. Extreme One-Day Precipitation Events in the Contiguous 48 States, 1910–2011

Figure 1 was developed as part of NOAA's CEI, an index that uses six different variables to examine trends in extreme weather and climate. This figure shows trends in the prevalence of extreme one-day precipitation events, based on a component of NOAA's CEI (labeled as Step 4) that looks at the percentage of land area within the contiguous 48 states that experienced a much greater than normal proportion of precipitation derived from extreme one-day precipitation events in any given year.

In compiling the CEI, NOAA applied more stringent criteria to select only those stations with data for at least 90 percent of the days in each year as well as 90 percent of the days during the full period of record. Applying these criteria resulted in the selection of only a subset of USHCN stations. To supplement the USHCN record, the CEI (and hence Figure 1) also includes data from NOAA's Cooperative Summary of the Day (TD3200) and pre-1948 (TD3206) daily precipitation stations. This resulted in a total of over 1,300 precipitation stations.

NOAA scientists computed the data for the CEI and calculated the percentage of land area for each year. They performed these steps by dividing the contiguous 48 states into a 1-degree by 1-degree grid and using data from one station per each grid box, rather than multiple stations. This was done to eliminate many of the artificial extremes that resulted from a changing number of available stations over time.

For each grid cell, the indicator looks at what portion of the total annual precipitation occurred on days that had "extreme" precipitation totals. Thus, the indicator essentially describes what percentage of precipitation is arriving in short, intense bursts. "Extreme" is defined as the highest 10th percentile, meaning an "extreme" one-day event is one in which the total precipitation received at a given location during the course of the day is at the upper end of the distribution of expected values (i.e., the distribution of all one-day precipitation totals at that location during the period of record). After extreme one-day events were identified, the percentage of annual precipitation occurring on extreme days was calculated for each year at each location. The subsequent step looked at the distribution of these percentage values over the full period of record, then identified all years that were in the highest 10th percentile. These years were considered to have a "greater than normal" amount of precipitation derived from extreme precipitation events at a given location. The top 10th percentile was chosen so as to give the overall index an expected value of 10 percent. Finally, data were aggregated nationwide to determine the percentage of land area with "greater than normal" precipitation derived from extreme events in each year.

The CEI can be calculated for individual seasons or for an entire year. This indicator uses the annual CEI, which is shown by the columns in Figure 1. To smooth out some of the year-to-year variability, EPA applied a nine-point binomial filter, which is plotted at the center of each nine-year window. For example, the smoothed value from 2002 to 2010 is plotted at year 2006. NOAA NCDC recommends this approach and has used it in the official online reporting tool for the CEI.

EPA used endpoint padding to extend the nine-year smoothed lines all the way to the ends of the period of record. As recommended by NCDC, EPA calculated smoothed values as follows: If 2011 was the most recent year with data available, EPA calculated smoothed values to be centered at 2008, 2009, 2010, and 2011 by inserting the 2011 data point into the equation in place of the as-yet-unreported annual data points for 2012 and beyond. EPA used an equivalent approach at the beginning of the time series.

The CEI has been extensively documented and refined over time to provide the best possible representation of trends in extreme weather and climate. For an overview of how NOAA constructed Step 4 of the CEI, see: www.ncdc.noaa.gov/oa/climate/research/cei/cei.html. This page provides a list of references that describe analytical methods in greater detail. In particular, see Gleason et al. (2008).

Figure 2. Unusually High Annual Precipitation in the Contiguous 48 States, 1895–2011

Figure 2 shows trends in the occurrence of abnormally high annual total precipitation based on the SPI, which is an index based on the probability of receiving a particular amount of precipitation in a given location. Thus, this index essentially compares the actual amount of annual precipitation received at a particular location with the amount that would be expected based on historical records. An SPI value of zero represents the median of the historical distribution; a negative SPI value represents a drier-than-normal period and a positive value represents a wetter-than-normal period.

The Western Regional Climate Center (WRCC) calculates the SPI by dividing the contiguous 48 states into 344 regions called "climate divisions" and analyzing data from weather stations within each division. A typical division has 10 to 50 stations, some from USHCN and others from the broader set of cooperative weather stations. For a given time period, WRCC calculated a single SPI value for each climate division based on an unweighted average of data from all stations within the division. This procedure has been followed for data from 1931 to present. A regression technique was used to compute divisional values prior to 1931 (Guttman and Quayle, 1996).

WRCC and NOAA calculate the SPI for various time periods ranging from one month to 24 months. This indicator uses the 12-month SPI data reported for the end of December of each year (1895 to 2011). The 12-month SPI is based on precipitation totals for the previous 12 months, so a December 12-month SPI value represents conditions over the full calendar year.

To create Figure 2, EPA identified all climate divisions with an SPI value of +2.0 or greater in a given year, where +2.0 is a suggested threshold for "abnormally high" precipitation (i.e., the upper tail of the historical distribution). For each year, EPA then determined what percentage of the total land area of the contiguous 48 states these "abnormally high" climate divisions represent. This annual percentage value is represented by the thin curve in the graph. To smooth out some of the year-to-year variability, EPA applied a nine-point binomial filter, which is plotted at the center of each nine-year window. For example, the smoothed value from 2002 to 2010 is plotted at year 2006. NOAA NCDC recommends this approach and has used it in the official online reporting tool for the CEI (the source of Figure 1).

EPA used endpoint padding to extend the nine-year smoothed lines all the way to the ends of the period of record. As recommended by NCDC, EPA calculated smoothed values as follows: If 2011 was the most recent year with data available, EPA calculated smoothed values to be centered at 2008, 2009, 2010, and 2011 by inserting the 2011 data point into the equation in place of the as-yet-unreported annual data points for 2012 and beyond. EPA used an equivalent approach at the beginning of the time series.

Like the CEI, the SPI is extensively documented in the peer-reviewed literature. The SPI is particularly useful among drought and precipitation indices because it can be applied over a variety of time frames and because it allows comparison of different locations and different seasons on a standard scale.

For an overview of the SPI and a list of resources describing methods used in constructing this index, see NDMC (2011) and the following websites:
http://lwf.ncdc.noaa.gov/oa/climate/research/prelim/drought/spi.html and
www.wrcc.dri.edu/spi/explanation.html. For more information on climate divisions and the averaging and regression processes used to generalize values within each division, see Guttman and Quayle (1996).

This indicator does not attempt to project data backward before the start of regular data collection or forward into the future. All values of the indicator are based on actual measured data. No attempt has been made to interpolate days with missing data. Rather, the issue of missing data was addressed in the site selection process by including only those stations that had very few missing data points.

7. Quality Assurance and Quality Control

USHCN precipitation data have undergone extensive quality assurance and quality control (QA/QC) procedures to identify errors and biases in the data and either remove these stations from the time series or apply correction factors. These quality control procedures are summarized at: www.ncdc.noaa.gov/oa/climate/research/ushcn/#processing. A series of data corrections was developed to specifically address potential problems in trend estimation in USHCN Version 2. They include:

- Removal of duplicate records
- Procedures to deal with missing data
- Testing and correcting for artificial discontinuities in a local station record, which might reflect station relocation or instrumentation changes

Data from weather stations also undergo routine QC checks before they are added to historical databases in their final form. These steps are typically performed within four months of data collection (NDMC, 2011).

QA/QC procedures are not readily available for the CEI and SPI, but both of these indices have been published in the peer-reviewed literature, indicating a certain degree of rigor.

Analysis

8. Comparability Over Time and Space

To be included in the USHCN, a station had to meet certain criteria for record longevity, data availability (percentage of missing values), spatial coverage, and consistency of location (i.e., experiencing few station changes). The period of record varies for each station but generally includes most of the 20th century. One of the objectives in establishing the USHCN was to detect secular changes of regional rather than local climate. Therefore, stations included in the network are only those believed to not be influenced to any substantial degree by artificial changes of local environments.

9. Sources of Uncertainty

Error estimates are not readily available for daily precipitation measurements or for the CEI and SPI calculations that appear in this indicator. In general, uncertainties in precipitation data increase as one goes back in time, as there are fewer stations early in the record. However, these uncertainties should not be sufficient to undermine the fundamental trends in the data. The USHCN has undergone extensive testing to identify errors and biases in the data and either remove these stations from the time series or apply scientifically appropriate correction factors to improve the utility of the data. In addition, both parts of the indicator have been restricted to stations meeting specific criteria for data availability.

10. Sources of Variability

Precipitation varies from location to location and from year to year as a result of normal variation in weather patterns, multi-year climate cycles such as the El Niño–Southern Oscillation and Pacific Decadal Oscillation, and other factors. This indicator accounts for these factors by presenting a long-term record (a century of data) and aggregating consistently over time and space.

11. Statistical/Trend Analysis

EPA has determined that the time series in Figure 1 has an increasing trend of approximately half a percentage point per decade and the time series in Figure 2 has an increasing trend of approximately 0.15 percentage points per decade. Both of these trends were calculated by ordinary least-squares regression, which is a common statistical technique for identifying a first-order trend. Analyzing the significance of these trends would potentially require consideration of serial correlation and other more complex statistical factors.

12. Data Limitations

Factors that may impact the confidence, application, or conclusions drawn from this indicator are as follows:

1. Both figures are national in scope, meaning they do not provide information about trends in extreme or heavy precipitation on a local or regional scale.
2. Weather monitoring stations tend to be closer together in the eastern and central states than in the western states. In areas with fewer monitoring stations, heavy precipitation indicators are less likely to reflect local conditions accurately.
3. The indicator does not include Alaska, which has seen some notable changes in heavy precipitation in recent years (e.g., Gleason et al., 2008).

References

Gleason, K.L., J.H. Lawrimore, D.H. Levinson, T.R. Karl, and D.J. Karoly. 2008. A revised U.S. climate extremes index. J. Climate 21:2124–2137.

Guttman, N.B., and R.G. Quayle. 1996. A historical perspective of U.S. climate divisions. Bull. Am. Meteorol. Soc. 77(2):293–303. www.ncdc.noaa.gov/oa/climate/research/cag3/i1520-0477-077-02-0293.pdf.

NDMC (National Drought Mitigation Center). 2011. Data source and methods used to compute the Standardized Precipitation Index. http://drought.unl.edu/MonitoringTools/ClimateDivisionSPI/DataSourceMethods.aspx.

Drought

Identification

1. Indicator Description

This indicator measures drought conditions of U.S. lands from 1895 to 2011. Drought conditions can affect agriculture, water supplies, energy production, and many other aspects of society.

Components of this indicator include:

- Average drought conditions in the contiguous 48 states over time, based on the Palmer Drought Severity Index (Figure 1)
- Percent of U.S. lands classified under drought conditions in recent years, based on an index called the U.S. Drought Monitor (Figure 2)

2. Revision History

April 2010: Indicator posted
December 2011: Updated with U.S. Drought Monitor data through 2010; added a new figure based on the Palmer Drought Severity Index (PDSI)
January 2012: Updated with data through 2011

Data Sources

3. Data Sources

Data for Figure 1 were obtained from the National Oceanic and Atmospheric Administration's (NOAA's) National Climatic Data Center (NCDC), which maintains a large collection of climate data online.

Data for Figure 2 were provided by the U.S. Drought Monitor. Historical data in table form are available at: http://droughtmonitor.unl.edu/archive.html. Maps and current drought information can be found on the main Drought Monitor website at: http://droughtmonitor.unl.edu.

4. Data Availability

Figure 1. Average Drought Conditions in the Contiguous 48 States, 1895–2011

NCDC provides access to monthly values of the PDSI averaged across the entire contiguous 48 states, which EPA downloaded for this indicator. These data are available at: http://www7.ncdc.noaa.gov/CDO/CDODivisionalSelect.jsp. This website also provides access to monthly PDSI values for nine broad regions, individual states, and 344 smaller regions called "climate divisions" (one to 10 climate divisions per state). For accompanying metadata, see: http://www7.ncdc.noaa.gov/CDO/DIV_DESC.txt.

PDSI values are calculated from precipitation and temperature measurements collected by weather stations within each climate division. Individual station measurements and metadata are available through NCDC's website (http://www.ncdc.noaa.gov/oa/ncdc.html).

Figure 2. U.S. Lands Under Drought Conditions, 2000–2011

U.S. Drought Monitor data can be obtained from: http://droughtmonitor.unl.edu/archive.html. Select "Tables" and "United States" to view the historical data that were used for this indicator. For each week, the data table shows what percentage of land area was under the following drought conditions:

1. None
2. D0–D4
3. D1–D4
4. D2–D4
5. D3–D4
6. D4 alone

This indicator covers the time period from 2000 to 2011. Although data were available for parts of 1999 and 2012 at the time EPA last updated this indicator, EPA chose to report only full years.

Drought Monitor data are based on a wide variety of underlying sources. Some are readily available from public websites; others might require specific database queries or assistance from the agencies that collect and/or compile the data. For links to many of the data sources, see: http://droughtmonitor.unl.edu/links.html.

Methodology

5. Data Collection

Figure 1. Average Drought Conditions in the Contiguous 48 States, 1895–2011

The PDSI is calculated from daily temperature measurements and precipitation totals collected at thousands of weather stations throughout the United States. These stations are overseen by NOAA, and they use standard instruments to measure temperature and precipitation. Some of these stations are first-order stations operated by NOAA's National Weather Service. The remainder are Cooperative Observer Program (COOP) stations operated by other organizations using trained observers and equipment and procedures prescribed by NOAA. For an inventory of U.S. weather stations and information about data collection methods, see: http://www.ncdc.noaa.gov/oa/land.html#dandp, www.ncdc.noaa.gov/oa/climate/research/ushcn, and the technical reports and peer-reviewed papers cited therein.

Figure 2. U.S. Lands Under Drought Conditions, 2000–2011

Figure 2 is based on the U.S. Drought Monitor, which uses a comprehensive definition of "drought" that accounts for a large number of different physical variables. Many of the underlying variables reflect weather and climate, including daily precipitation totals collected at weather stations throughout the United States as described above for Figure 1. Other parameters include measurements of soil moisture, streamflow, reservoir and groundwater levels, and vegetation health. These measurements are generally collected by government agencies following standard methods, such as a national network of stream gauges that measure daily (and weekly) flow, comprehensive satellite mapping programs, and other systematic monitoring networks. Each program has its own sampling or monitoring design. The Drought Monitor and the other drought indices that contribute to it have been formulated such that they rely on measurements that offer sufficient temporal and spatial resolution.

The U.S. Drought Monitor has five primary inputs:

- The PDSI
- The Soil Moisture Model, from NOAA's Climate Prediction Center
- Weekly streamflow data from the U.S. Geological Survey
- The Standardized Precipitation Index (SPI), compiled by NOAA and the Western Regional Climate Center (WRCC)
- A blend of objective short- and long-term drought indicators (short-term drought indicator blends focus on 1- to 3-month precipitation totals; long-term blends focus on 6 to 60 months)

At certain times and in certain locations, the Drought Monitor also incorporates one or more of the following additional indices, some of which are particularly well-suited to the growing season and others of which are ideal for snowy areas or ideal for the arid West:

- A topsoil moisture index from the U.S. Department of Agriculture's National Agricultural Statistics Service
- The Keetch-Byram Drought Index
- Vegetation health indices based on satellite imagery from NOAA's National Environmental Satellite, Data, and Information Service (NESDIS)
- Snow water content
- River basin precipitation
- The Surface Water Supply Index (SWSI)
- Groundwater levels
- Reservoir storage
- Pasture or range conditions

For more information on the other drought indices that contribute to the Drought Monitor, including the data used as inputs to these other indices, see: http://drought.unl.edu/Planning/Monitoring/ComparisonofIndicesIntro.aspx.

To find information on underlying sampling methods and procedures for constructing some of the component indices that go into determining the U.S. Drought Monitor, one will need to consult a variety of additional sources. For example, as described above for Figure 1, NCDC has published extensive documentation about methods for collecting precipitation data.

6. Indicator Derivation

Figure 1. Average Drought Conditions in the Contiguous 48 States, 1895–2011

PDSI calculations are designed to reflect the amount of moisture available at a particular location and point in time, based on the amount of precipitation received as well as the temperature, which influences evaporation rates. The formula for creating this index was originally proposed in the 1960s (Palmer, 1965). Since then, the methods have been tested extensively and used to support hundreds of published studies. The PDSI is the most widespread and scientifically vetted drought index in use today.

The PDSI was designed to characterize long-term drought (i.e., patterns lasting a month or more). Because drought is cumulative, the formula takes precipitation and temperature data from previous weeks and months into account. Thus, a single rainy day is unlikely to cause a dramatic shift in the index.

PDSI values are normalized relative to long-term average conditions at each location, which means this method can be applied to any location regardless of how wet or dry it typically is. NOAA currently uses 1931–1990 as its long-term baseline. The index essentially measures deviation from normal conditions. The PDSI takes the form of a numerical value, generally ranging from -6 to +6. A value of zero reflects average conditions. Negative values indicate drier-than-average conditions and positive values indicate wetter-than-average conditions. NOAA provides the following interpretations for specific ranges of the index:

- 0 to -0.5 = normal
- -0.5 to -1.0 = incipient drought
- -1.0 to -2.0 = mild drought
- -2.0 to -3.0 = moderate drought
- -3.0 to -4.0 = severe drought
- < -4.0 = extreme drought

Similar adjectives can be applied to positive (wet) values.

NOAA calculates monthly values of the PDSI for each of the 344 climate divisions within the contiguous 48 states. These values are calculated from weather stations reporting both temperature and precipitation. All stations within a division are given equal weight. NOAA also combines PDSI values from all climate divisions to derive a national average for every month.

EPA obtained monthly national monthly PDSI values from NOAA, then calculated annual averages. To smooth out some of the year-to-year variability, EPA applied a nine-point binomial filter, which is plotted at the center of each nine-year window. For example, the smoothed value from 2002 to 2010 is plotted at year 2006. NOAA NCDC recommends this approach. Figure 1 shows both the annual values and the smoothed curve.

EPA used endpoint padding to extend the nine-year smoothed lines all the way to the ends of the period of record. As recommended by NCDC, EPA calculated smoothed values as follows: If 2011 was the most recent year with data available, EPA calculated smoothed values to be centered at 2008, 2009, 2010, and 2011 by inserting the 2011 data point into the equation in place of the as-yet-unreported annual data points for 2012 and beyond. EPA used an equivalent approach at the beginning of the time series.

For more information about NOAA's processing methods, see the metadata file at: http://www7.ncdc.noaa.gov/CDO/DIV_DESC.txt. NOAA's website provides a variety of other references regarding the PDSI at: http://www.ncdc.noaa.gov/oa/climate/research/prelim/drought/palmer.html.

Figure 2. U.S. Lands Under Drought Conditions, 2000–2011

The National Drought Mitigation Center at the University of Nebraska–Lincoln produces the U.S. Drought Monitor with assistance from many other climate and water experts at the federal, regional, state, and local levels. For each week, the Drought Monitor labels areas of the country according to the intensity of any drought conditions that may be present. An area experiencing drought is assigned a score ranging from D0, the least severe drought, to D4, the most severe. For definitions of these classifications, see: http://droughtmonitor.unl.edu/classify.htm.

Drought Monitor values are determined from the five major components and other supplementary factors listed in Section 5. A table on the Drought Monitor website (http://droughtmonitor.unl.edu/classify.htm) explains the range of observed values for each major component that would result in a particular Drought Monitor score. The final index score is based to some degree on expert judgment, however. For example, expert analysts resolve discrepancies in cases where the five major components might not coincide with one another. They might assign a final Drought Monitor score based on what the majority of the components suggest, or they might weight the components differently according to how well they perform in various parts of the country and at different times of the year. Experts also determine what additional factors to consider for a given time and place and how heavily to weight these supplemental factors. For example, snowpack is particularly important in the West, where it has a strong bearing on water supplies.

From the Drought Monitor's public website, EPA obtained data covering the contiguous 48 states plus Alaska, Hawaii, and Puerto Rico, then performed a few additional calculation steps. The original data set reports cumulative categories (for example, "D2–D4" and "D3–D4"), so EPA had to subtract one category from another in order to find the percentage of land area belonging to each individual drought category (e.g., D2 alone). EPA also calculated annual averages to support some of the statements presented in the "Key Points" for this indicator.

No attempt has been made to portray data outside the time and space where measurements were made. Measurements are collected on at least a weekly basis (in the case of some variables like precipitation and streamflow, at least daily) and used to derive weekly maps for the U.S. Drought Monitor. Values are generalized over space by weighting the different factors that go into calculating the overall index and applying expert judgment to derive the final weekly map and the corresponding totals for affected area.

For more information about how the Drought Monitor is calculated, including percentiles associated with the occurrence of each of the D0–D4 classifications, see Svoboda et al. (2002) along with the documentation provided on the Drought Monitor website at: http://droughtmonitor.unl.edu.

7. Quality Assurance and Quality Control

Figure 1. Average Drought Conditions in the Contiguous 48 States, 1895–2011

Data from weather stations go through a variety of quality assurance and quality control (QA/QC) procedures before they can be added to historical databases in their final form. NOAA's U.S. Historical Climatology Network—one of the main weather station databases—follows strict QA/QC procedures to identify errors and biases in the data and either remove these stations from the time series or apply correction factors. Procedures for the USHCN are summarized at: www.ncdc.noaa.gov/oa/climate/research/ushcn/#processing. Specific to this indicator, NOAA's metadata file (http://www7.ncdc.noaa.gov/CDO/DIV_DESC.txt) and Karl et al. (1986) describe steps that have been taken to reduce biases associated with differences in the time of day when temperature observations are reported.

Figure 2. U.S. Lands Under Drought Conditions, 2000–2011

QA/QC procedures for the overall U.S. Drought Monitor data set are not readily available. Each underlying data source has its own methodology, which typically includes some degree of QA/QC. For example, precipitation and temperature data are verified and corrected as described above for Figure 1. Some of the other underlying data sources have QA/QC procedures available online, but others do not.

Analysis

8. Comparability Over Time and Space

Figure 1. Average Drought Conditions in the Contiguous 48 States, 1895–2011

PDSI calculation methods have been applied consistently over time and space. In all cases, the index relies on the same underlying measurements (precipitation and temperature). Although fewer stations were collecting weather data during the first few decades of the analysis, NOAA has determined that enough stations were available starting in 1895 to calculate valid index values for the contiguous 48 states as a whole.

Figure 2. U.S. Lands Under Drought Conditions, 2000–2011

The resolution of the U.S. Drought Monitor has improved over time. When the Drought Monitor began to be calculated in 1999, many of the component indicators used to determine drought conditions were reported at the climate division level. Many of these component indicators now include data from the county and sub-county level. This change in resolution over time can be seen in the methods used to draw contour lines on Drought Monitor maps.

The drought classification scheme used for this indicator is produced by combining data from several different sources. Different locations may use different primary sources—or the same sources, weighted differently. These data are combined to reflect the collective judgment of experts and in some cases are adjusted to reconcile conflicting trends shown by different data sources over different time periods.

Though data resolution and mapping procedures have varied somewhat over time and space, the fundamental construction of the indicator has remained consistent.

9. Sources of Uncertainty

Error estimates are not readily available for national average PDSI, the U.S. Drought Monitor, or the underlying measurements that contribute to this indicator. It is not clear how much uncertainty might be associated with the component indices that go into formulating the Drought Monitor or the process of compiling these indices into a single set of weekly values through averaging, weighting, and expert judgment.

10. Sources of Variability

Conditions associated with drought naturally vary from place to place and from one day to the next, depending on weather patterns and other factors. Both figures address spatial variability by presenting aggregate national trends. Figure 1 addresses temporal variability by using an index that is designed to measure long-term drought and is not easily swayed by short-term conditions. Figure 1 also provides an annual average, along with a nine-year smoothed average. Figure 2 smoothes out some of the inherent variability in drought measurement by relying on many indices, including several with a long-term focus. While Figure 2 shows noticeable week-to-week variability, it also reveals larger year-to-year patterns.

11. Statistical/Trend Analysis

This indicator does not report on the slope of the trend in PDSI values over time, nor does it calculate the statistical significance of this trend.

Because data from the U.S. Drought Monitor are only available for the most recent decade, this metric is too short-lived to be used for assessing long-term climate trends. Furthermore, there is no clear long-term trend in Figure 2. With continued data collection, future versions of this indicator should be able to paint a more statistically robust picture of long-term trends in Drought Monitor values.

12. Data Limitations

Factors that may impact the confidence, application, or conclusions drawn from this indicator are as follows:

1. The indicator gives a broad overview of drought conditions in the United States. It is not intended to replace local or state information that might describe conditions more precisely for a particular region. Local or state entities might monitor different variables to meet specific needs or to address local problems. As a consequence, there could be water shortages or crop failures within an area not designated as a drought area, just as there could be locations with adequate water supplies in an area designated as D3 or D4 (extreme or exceptional) drought.
2. Because this indicator focuses on national trends, it does not show how drought conditions vary by region. For example, even if half of the country suffered from severe drought, Figure 1 could show an average index value close to zero if the rest of the country was wetter than average. Thus, Figure 1 might understate the degree to which droughts are becoming more severe in some areas while other places receive more rain as a result of climate change.

3. Although the PDSI is arguably the most widely used drought index, it has some limitations that have been documented extensively in the literature. While the use of just two variables (precipitation and temperature) makes this index relatively easy to calculate over time and space, drought can have many other dimensions that these two variables do not fully capture. For example, the PDSI loses accuracy in areas where a substantial portion of the water supply comes from snowpack.

4. Indices such as the U.S. Drought Monitor seek to address the limitations of the PDSI by incorporating many more variables. However, the Drought Monitor is relatively new and cannot yet be used to assess long-term climate trends.

5. The drought classification scheme used for Figure 2 is produced by combining data from several different sources. These data are combined to reflect the collective judgment of experts and in some cases are adjusted to reconcile conflicting trends shown by different data sources over different time periods.

References

Karl, T.R., C.N. Williams, Jr., P.J. Young, and W.M. Wendland. 1986. A model to estimate the time of observation bias associated with monthly mean maximum, minimum, and mean temperatures for the Unites States. Journal of Climate and Applied Meteorology 25(1).

Palmer, W.C. 1965. Meteorological drought. Res. Paper No.45. Washington, DC: U.S. Department of Commerce.

Svoboda, M., D. Lecomte, M. Hayes, R. Heim, K. Gleason, J. Angel, B. Rippey, R. Tinker, M. Palecki, D. Stooksbury, D. Miskus, and S. Stephens. 2002. The drought monitor. Bull. Am. Meteorol. Soc. 83(8):1181–1190.

Tropical Cyclone Activity

Identification

1. Indicator Description

This indicator examines the aggregate activity of hurricanes and other tropical storms in the Atlantic Ocean, Caribbean, and Gulf of Mexico between 1949 and 2011. Climate change is expected to affect tropical cyclone activity through increased sea surface temperatures and other environmental changes that are key influences on cyclone formation and behavior.

Components of this indicator include:

- The number of hurricanes in the North Atlantic each year, along with the number making landfall in the United States (Figure 1)
- Frequency, intensity, and duration of North Atlantic cyclones as measured by the Accumulated Cyclone Energy Index (Figure 2)
- Frequency, intensity, and duration of North Atlantic cyclones as measured by the Power Dissipation Index (Figure 3)

2. Revision History

April 2010: Indicator posted
December 2011: Updated Figure 2 with data through 2011
April 2012: Added hurricane counts (new Figure 1)
May 2012: Updated Figure 3 with data through 2011

Data Sources

3. Data Sources

This indicator is based on data maintained by the National Oceanic and Atmospheric Administration's (NOAA's) National Hurricane Center in a database referred to as HURDAT (HURricane DATa). This indicator presents three separate analyses of HURDAT data: a set of hurricane counts compiled by NOAA, NOAA's Accumulated Cyclone Energy (ACE) Index, and the Power Dissipation Index (PDI) developed by Dr. Kerry Emanuel at the Massachusetts Institute of Technology (MIT).

4. Data Availability

Figure 1. Number of Hurricanes in the North Atlantic, 1878–2011

Data for Figure 1 were provided by Tom Knutson at NOAA, based on a compilation of several published datasets:

- A list of U.S. landfalling hurricanes maintained by NOAA's Atlantic Oceanographic and Meteorological Laboratory (AOML), Hurricane Research Division (HRD): www.aoml.noaa.gov/hrd/hurdat/All_U.S._Hurricanes.html.
- Total hurricane counts maintained by NOAA HRD at: www.aoml.noaa.gov/hrd/hurdat/tracks1851to2011_atl_reanal.html.
- Raw and adjusted total hurricane counts through 2010, posted by NOAA at: www.gfdl.noaa.gov/index/cms-filesystem-action/user_files/gav/historical_storms/vk_11_hurricane_counts.txt (linked from: www.gfdl.noaa.gov/gabriel-vecchi-noaa-gfdl).

Figure 2. North Atlantic Cyclone Intensity According to the Accumulated Cyclone Energy Index, 1950–2011

An overview of the ACE Index is available at: www.cpc.ncep.noaa.gov/products/outlooks/background_information.shtml. The data for this indicator are published in the form of a bar graph in NOAA's annual "North Atlantic Hurricane Season: A Climate Perspective" (2011 edition available at: www.cpc.ncep.noaa.gov/products/expert_assessment/hurrsummary_2011.pdf). The numbers were obtained in spreadsheet form by contacting Dr. Gerry Bell at NOAA.

Figure 3. North Atlantic Cyclone Intensity According to the Power Dissipation Index, 1949–2011

Emanuel (2005, 2007) gives an overview of the PDI, along with figures and tables. This indicator reports on an updated version of the data set (through 2011) that was provided by Dr. Kerry Emanuel.

Underlying Data

Wind speed measurements and other HURDAT data are available in various formats on NOAA's AOML website: www.aoml.noaa.gov/hrd/hurdat/Data_Storm.html. Some documentation is available at: www.aoml.noaa.gov/hrd/hurdat/Documentation.html, and definitions for HURDAT data formats are available at: www.aoml.noaa.gov/hrd/data_sub/hurdat.html.

Methodology

5. Data Collection

This indicator is based on measurements of tropical cyclones over time. HURDAT compiles information on all hurricanes and other tropical storms occurring in the North Atlantic Ocean, including parameters such as wind speed, barometric pressure, storm tracks, and dates. Field methods for data collection and analysis are documented in official NOAA publications (Jarvinen et al., 1984). This indicator is based on sustained wind speed, which is defined as the one-minute average wind speed at an altitude of 10 meters.

Data collection methods have evolved over time. When data collection began, ships and land observation stations were used to measure and track storms. Analysts compiled all available wind speed observations and all information about the measurement technique to determine the wind speed for the four daily intervals for which the storm track was recorded.

More recently, organized aircraft reconnaissance, the coastal radar network, and weather satellites with visible and infrared sensors have improved accuracy in determining storm track, maximum wind speeds, and other storm parameters such as central pressure. Weather satellites were first used in the 1960s to detect the initial position of a storm system; reconnaissance aircraft would then fly to the location to collect precise measurements of the wind field, central pressure, and location of the center. Data collection methods have since improved with more sophisticated satellites.

This indicator covers storms occurring in the Atlantic Ocean north of the equator, including the Caribbean Sea and the Gulf of Mexico. HURDAT does not include data for storm systems that are classified as extratropical. However, it does include data from storms classified as subtropical, meaning they exhibit some characteristics of a tropical cyclone but also some characteristics of an extratropical storm. Subtropical cyclones are now named in conjunction with the tropical storm naming scheme, and in practice, many subtropical storms eventually turn into tropical storms. HURDAT is updated annually by NOAA and data are available from 1886 through 2011.

Sampling and analysis procedures for the HURDAT data are described by Jarvinen et al. (1984) for collection methods up to 1984. Changes to past collection methods are partially described in the supplementary methods from Emanuel (2005). Other data explanations are available at: www.nhc.noaa.gov/pastall.shtml#hurdat. The mission catalogue of data sets collected by NOAA aircraft is available at: www.aoml.noaa.gov/hrd/data_sub/hurr.html.

6. Indicator Derivation

Figure 1. Number of Hurricanes in the North Atlantic, 1878–2011

This figure displays three time series: the number of hurricanes per year making landfall in the United States, the total number of hurricanes on record for the North Atlantic, and an adjusted total that attempts to account for changes in observing capabilities. All three counts are limited to cyclones in the North Atlantic (i.e., north of the equator) meeting the definition of a hurricane, which requires sustained wind speeds of at least 74 miles per hour.

Landfalling counts reflect the following considerations:

- If a single hurricane made multiple U.S. landfalls, it is only counted once.
- If the hurricane center did not make a U.S. landfall (or substantially weakened before making landfall), but did produce hurricane-force winds over land, it is counted.
- If the hurricane center made landfall in Mexico, but did produce hurricane-force winds over the United States, it is counted.

For all years prior to the onset of complete satellite coverage in 1966, total basin-wide counts have been adjusted upward based on historical records of ship track density. In other words, during years when fewer ships that were making observations in a given ocean region, hurricanes in that region were more likely to have been missed, or their intensity underestimated to be below hurricane strength, leading to a larger corresponding adjustment to the count for those years. These adjustment methods are cited in Knutson et al. (2010) and described in more detail by Vecchi and Knutson (2008), Landsea et al. (2010), and Vecchi and Knutson (2011).

All three curves have been smoothed using a five-year unweighted average, as recommended by the data provider. Data are plotted at the center of each window; for example, the five-year smoothed value for 1949 to 1953 is plotted at year 1951. Because of this smoothing procedure and the absence of endpoint padding, no averages can be plotted for the first two years and last two years of the period of record (1878, 1879, 2010, and 2011).

Figure 2. North Atlantic Tropical Cyclone Activity According to the Accumulated Cyclone Energy Index, 1950–2011

This figure uses NOAA's ACE Index to describe the combined frequency, strength, and duration of tropical storms and hurricanes each season. As described by Bell and Chelliah (2006), "the ACE Index is calculated by summing the squares of the estimated 6-hourly maximum sustained wind speed in knots for all periods while the system is either a tropical storm or hurricane." A system is considered at least a tropical storm if it has a wind speed of at least 39 miles per hour. The ACE Index is preferred over other similar indices such as the Hurricane Destruction Potential (HDP) and the Net Tropical Cyclone Index (NTC) because it takes tropical storms into account and it does not include multiple sampling of some parameters. The ACE Index also includes subtropical cyclones, which are named using the same scheme as tropical cyclones and may eventually turn into tropical cyclones in some cases. The index does not include information on storm size, which is an important component of a storm's damage potential.

Figure 2 of the indicator shows annual values of the ACE, which are determined by summing the individual ACE Index values of all storms during that year. The index itself is measured in units of wind speed squared, but for this indicator, the index has been converted to a numerical scale where 100 equals the median value over a base period from 1981 to 2010. A value of 150 would therefore represent 150 percent of the median, or 50 percent more than normal. NOAA has also established a set of thresholds to categorize each hurricane season as "above normal," "near normal," or "below normal" based on the distribution of observed values during the base period. The "near normal" range extends from 71.5 to 111 percent of the median, with the "above normal" range above 111 percent of the median and the "below normal" range below 71.5 percent.

ACE Index computation methods and seasonal classifications are described by Bell and Chelliah (2006). This information is also available on the NOAA website at: www.cpc.noaa.gov/products/outlooks/background_information.shtml.

Figure 3. North Atlantic Tropical Cyclone Activity According to the Power Dissipation Index, 1949–2011

For additional perspective, this figure presents the PDI. Like the ACE Index, the PDI is also based on storm frequency, wind speed, and duration, but it uses a different calculation method that places more emphasis on storm intensity by using the cube of the wind speed rather than the wind speed squared (as for the ACE). Emanuel (2005, 2007) provides a complete description of how the PDI is calculated. Emanuel (2007) also explains adjustments that were made to correct for biases in the quality of storm observations and wind speed measurements early in the period of record. The PDI data in Figure 3 of this indicator are in units of 10^{11} m^3/s^2, but the actual figure omits this unit and simply alludes to "index values" in order to make the indicator accessible to the broadest possible audience.

The PDI data shown in Figure 3 have been smoothed using a five-year weighted average applied with weights of 1, 3, 4, 3, and 1. This method applies greater weight to values near the center of each five-year window. Data are plotted at the center of each window; for example, the five-year smoothed value for 1949 to 1953 is plotted at year 1951. The data providers recommend against endpoint padding for these particular variables, based on past experience and their expert judgment, so no averages can be plotted for the first two years and last two years of the period of record (1949, 1950, 2010, and 2011).

The PDI includes all storms that are in the so-called "best track" data set issued by NOAA, which can include subtropical storms. Weak storms contribute very little to power dissipation, however, so subtropical storms typically have little impact on the final metric.

Emanuel (2005, 2007) describes methods for calculating the PDI and deriving the underlying power dissipation formulas. Analysis techniques, data sources, and corrections to raw data used to compute the PDI are described in the supplementary methods for Emanuel (2005), with further corrections addressed in Emanuel (2007).

Sea surface temperature has been plotted for reference, based on methods described in Emanuel (2005, 2007). The curve in Figure 3 represents average sea surface temperature in the area of storm genesis in the North Atlantic: specifically, a box bounded in latitude by 6°N and 18°N, and in longitude by 20°W and 60°W. Values have been smoothed over five-year periods. For the sake of straightforward presentation, sea surface temperature has been plotted in unitless form without a secondary axis, and the curve has been positioned to clearly show the relationship between sea surface temperature and the PDI.

7. Quality Assurance and Quality Control

Jarvinen et al. (1984) describe quality assurance/quality control procedures for each of the variables in the HURDAT data set. Corrections to early HURDAT data are made on an ongoing basis through the HURDAT re-analysis project to correct for both systematic and random errors identified in the data set. Information on this re-analysis is available at on the NOAA website at: www.aoml.noaa.gov/hrd/data_sub/re_anal.html. Emanuel (2005) provides a "supplementary methods" document that describes both the evolution of more accurate sample collection technology and further corrections made to the data.

Analysis

8. Comparability Over Time and Space

In the early years of the data set there is a high likelihood that some tropical storms went undetected, as observations of storms were made only by ships at sea and land-based stations. Storm detection improved over time as ship track density increased, and beginning in 1944 with the use of organized aircraft reconnaissance (Jarvinen et al., 1984). However, it was not until the late 1960s and beyond, when satellite coverage was generally available, that the Atlantic tropical cyclone frequency record can be assumed to be relatively complete. Because of the greater uncertainties inherent in earlier data, Figure 1 adjusts pre-1966 data to account for the density of ship observations, while Figures 2 and 3 exclude data prior to 1950 and 1949, respectively. In addition, a re-analysis of early HURDAT data (www.aoml.noaa.gov/hrd/data_sub/re_anal.html) was initiated to improve both random and systematic error present in data from the beginning of the time series.

Emanuel (2005) describes the evolution of more accurate sample collection technology and various corrections made to the data. For the PDI, Emanuel (2007) employed an additional bias correction process for the early part of the period of record (the 1950s and 1960s), when aircraft reconnaissance and radar technology were less robust than they are today—possibly resulting in missed storms or underestimated power. These additional corrections were prompted in part by an analysis published by Landsea (1993).

9. Sources of Uncertainty

Counts of landfalling U.S. hurricanes are considered reliable back to the late 1800s, as population centers and recordkeeping were present all along the Gulf and Atlantic coasts at the time. Total hurricane counts for the North Atlantic became fairly reliable after aircraft reconnaissance began in 1944, and became highly reliable after the onset of satellite tracking around 1966. Prior to the use of these two methods, however, detection of non-landfalling storms depended on observations from ships, which could lead to undercounting due to low density of ship coverage. Figure 1 shows how pre-1966 counts have been adjusted upward based on the density of ship tracks (Vecchi and Knutson, 2011).

The ACE Index and the PDI are calculated directly from wind speed measurements. Thus, the main source of possible uncertainty in the indicator is uncertainties within the underlying HURDAT data set. Uncertainty measurements do not appear to be readily available for HURDAT data. Because the determination of storm track and wind speed requires some expert judgment by analysts, some uncertainty is likely. Methodological improvements suggest that recent data may be somewhat more accurate than earlier measurements.

Because uncertainty varies depending on observation method, and these methods have evolved over time, it is difficult to make a definitive statement about the impact of uncertainty on Figures 2 and 3. Changes in data gathering technologies could substantially influence the overall patterns in Figures 2 and 3, and the effects of these changes on data consistency over the life of the indicator would benefit from additional research.

10. Sources of Variability

Intensity varies by storm and location. The indicator addresses this type of variability by using two indices that aggregate all North Atlantic storms within a given year. Aggregate annual intensity also varies from year to year as a result of normal variation in weather patterns, multi-year climate cycles, and other factors. Annual storm counts can vary from year to year for similar reasons. Figure 2 shows interannual variability. Figures 1 and 3 also show variability over time, but they seek to focus on longer-term variability and trends by presenting a five-year smoothed curve.

Overall, it remains uncertain whether past changes in any tropical cyclone activity (frequency, intensity, rainfall, and so on) exceed the variability expected through natural causes, after accounting for changes over time in observing capabilities (Knutson et al., 2010).

11. Statistical/Trend Analysis

This indicator does not report on the slope of the apparent trends in hurricane counts or cyclone intensity, nor does it calculate the statistical significance of these trends. See Vecchi and Knutson (2008, 2011) for examples of such a trend analysis, including statistical significance tests.

12. Data Limitations

Factors that may impact the confidence, application, or conclusions drawn from this indicator are as follows:

1. Methods of detecting hurricanes have improved over time, and raw counts prior to the 1960s may undercount the total number of hurricanes that formed each year. However, Figure 1 presents an adjusted time series to attempt to address this limitation.
2. Wind speeds are measured using several observation methods with varying levels of uncertainty, and these methods have improved over time. The wind speeds recorded in HURDAT should be considered the best estimate of several wind speed observations compiled by analysts.
3. Many different indices have been developed to analyze storm duration, intensity, and threat. Each index has strengths and weaknesses associated with its ability to describe these parameters. The indices used in this indicator (hurricane counts, ACE Index, and PDI) are considered to be among the most reliable.

References

Bell, G.D., and M. Chelliah. 2006. Leading tropical modes associated with interannual and multidecadal fluctuations in North Atlantic hurricane activity. J. Climate 19:590–612.

Emanuel, K. 2005. Increasing destructiveness of tropical cyclones over the past 30 years. Nature 436:686–688. Supplementary methods available with the online version of the paper at: http://www.nature.com/nature/journal/v436/n7051/suppinfo/nature03906.html.

Emanuel, K. 2007. Environmental factors affecting tropical cyclone power dissipation. J. Climate 20(22):5497–5509. ftp://texmex.mit.edu/pub/emanuel/PAPERS/Factors.pdf.

Jarvinen, B.R., C.J. Neumann, and M.A.S. Davis. 1984. A tropical cyclone data tape for the North Atlantic Basin, 1886–1983: Contents, limitations and uses. NOAA Technical Memo NWS NHC 22.

Knutson, T.R., J.L. McBride, J. Chan, K. Emanuel, G. Holland, C. Landsea, I. Held, J.P. Kossin, A.K. Srivastava, and M. Sugi. 2010. Tropical cyclones and climate change. Nature Geosci. 3: 157–163.

Landsea, C.W. 1993. A climatology of intense (or major) Atlantic hurricanes. Mon. Weather Rev. 121:1703–1713. http://www.aoml.noaa.gov/hrd/Landsea/Landsea_MWRJune1993.pdf.

Landsea, C., G.A. Vecchi, L. Bengtsson, and T.R. Knutson. 2010. Impact of duration thresholds on Atlantic tropical cyclone counts. J. Climate 23:2508–2519. www.nhc.noaa.gov/pdf/landsea-et-al-jclim2010.pdf.

Vecchi, G.A., and T.R. Knutson. 2008. On estimates of historical North Atlantic tropical cyclone activity. J. Climate 21:3580–3600.

Vecchi, G.A., and T.R. Knutson. 2011. Estimating annual numbers of Atlantic hurricanes missing from the HURDAT database (1878–1965) using ship track density. J. Climate 24(6):1736–1746. http://www.gfdl.noaa.gov/bibliography/related_files/gav_2010JCLI3810.pdf.

Ocean Heat

Identification

1. Indicator Description

This indicator describes trends in the amount of heat stored in the world's oceans between 1955 and 2011. The amount of heat in the ocean, or ocean heat content, plays an important role in the Earth's climate system.

2. Revision History

April 2010: Indicator posted
April 2012: Updated with data through 2011

Data Sources

3. Data Sources

This indicator is based on analyses conducted by three different government agencies:
- Australia's Commonwealth Scientific and Industrial Research Organisation (CSIRO)
- Japan Agency for Marine-Earth Science and Technology (JAMSTEC)
- National Oceanic and Atmospheric Administration (NOAA)

JAMSTEC used four different datasets: the World Ocean Database (WOD), the World Ocean Atlas (WOA), the Global Temperature-Salinity Profile Program (GTSPP) (which was used to fill gaps in the WOD since 1990), and data from the Japan Maritime Self-Defense Force (JMSDF). CSIRO used two datasets: ocean temperature profiles in the ENACT/ENSEMBLES version 3 (EN3) and data collected using 60,000 Argo profiling floats. Additionally, CSIRO included bias-corrected Argo data, as described in Barker et al. (2011) and bias-corrected expendable bathythermograph (XBT) data from Wijffels et al. (2008). NOAA also used data from the WOD and WOA.

4. Data Availability

EPA created Figure 1 using trend data from three ongoing studies. Data and documentation from these studies can be found at the following links:

- CSIRO: www.cmar.csiro.au/sealevel/sl_data_cmar.html. Select "Updated Thermosteric Sea Level and Ocean Heat Content time series for 1950 to 2009" to download the data. See additional documentation in Domingues et al. (2008).
- JAMEST: Data from Ishii and Kimoto (2009) are posted at: http://atm-phys.nies.go.jp/~ism/pub/ProjD/doc. Updated data were provided by the author, Masayoshi Ishii. Data are expected to be updated regularly online in the future. See additional documentation in Ishii and Kimoto (2009).

- NOAA: www.nodc.noaa.gov/OC5/3M_HEAT_CONTENT. Select "basin time series," then under "yearly heat content," select "world." Use the "Yearly world 0-700 meters" file. See additional documentation in Levitus et al. (2009).

The underlying data for this indicator come from a variety of sources. Some of these datasets are publicly available, but other datasets consist of samples gathered by the authors of the source papers, and these data might be more difficult to obtain online. WOA and WOD data and descriptions of data are available on NOAA's National Oceanographic Data Center (NODC) website at: www.nodc.noaa.gov.

Methodology

5. Data Collection

This indicator reports on the amount of heat stored in the ocean from sea level to a depth of 700 meters, which accounts for approximately 17.5 percent of the total global ocean volume (calculation from Catia Domingues, CSIRO). Each of the three studies used to develop this indicator uses several ocean temperature profile datasets to calculate an ocean heat content trend line.

Several different devices are used to sample temperature profiles in the ocean. Primary methods used to collect data for this indicator include XBT; mechanical bathythermographs (MBT); Argo profiling floats; reversing thermometers; and conductivity, temperature, and depth sensors (CTD). These instruments produce temperature profile measurements of the ocean water column by recording data on temperature and depth. The exact methods used to record temperature and depth vary. For instance, XBTs use a fall rate equation to determine depth, whereas other devices measure depth directly.

More information on the three main studies and their respective methods can be found at:

- CSIRO: Domingues et al. (2008) and: www.cmar.csiro.au/sealevel/sl_data_cmar.html.
- JAMEST: Ishii and Kimoto (2009) and: http://atm-phys.nies.go.jp/~ism/pub/ProjD/doc.
- NOAA: Levitus et al. (2009) and: www.nodc.noaa.gov/OC5/3M_HEAT_CONTENT.

Studies that measure ocean temperature profiles are generally designed using in situ oceanographic observations and analyzed over a defined and spatially uniform grid (Ishii and Kimoto, 2009). For instance, the WOA dataset consists of in situ measurements of climatological fields, including temperature, measured in a 1-degree grid. Sampling procedures for WOD and WOA data are provided by NOAA's NODC at: www.nodc.noaa.gov/OC5/indprod.html. More information on the WOA sample design in particular can be found at: www.nodc.noaa.gov/OC5/WOA05/pr_woa05.html.

At the time of last update, CSIRO data were available through 2009, while data from the other two sources were available through 2011.

6. Indicator Derivation

While details of data analysis are particular to the individual study, in general, temperature profile data were averaged monthly at specific depths within rectangular grid cells. In some cases, interpolation techniques were used to fill gaps where observational spatial coverage was sparse. Additional steps were taken to correct for known biases in XBT data. Finally, temperature observations were used to calculate ocean heat content through various conversions. The model used to transform measurements was consistent across all three studies cited by this indicator.

Barker et al. (2011) describe instrument biases and procedures for correcting for these biases. For more information about interpolation and other analytical steps, see Ishii and Kimoto (2009), Domingues et al. (2008), Levitus et al. (2009), and references therein.

Each study used a different long-term average as a baseline. To allow more consistent comparison, EPA adjusted each curve such that its 1971–2000 average would be set at zero. Choosing a different baseline period would not change the shape of the data over time. Although some of the studies had pre-1955 data, Figure 1 begins at 1955 for consistency.

7. Quality Assurance and Quality Control

Data collection and archival steps included QA/QC procedures. For example, QA/QC measures for the WOA are available at: ftp://ftp.nodc.noaa.gov/pub/data.nodc/woa/PUBLICATIONS/qc94tso.pdf. Each of the data collection techniques involves different QA/QC measures. For example, a summary of studies concerning QA/QC of XBT data is available from NODC at: www.nodc.noaa.gov/OC5/XBT_BIAS/xbt_bibliography.html. The same site also provides additional information about QA/QC of ocean heat data made available by NODC.

All of the analyses performed for this indicator included additional QA/QC steps at the analytical stage. In each of the three main studies used in this indicator, the author carefully describes QA/QC methods or provides the relevant references.

Analysis

8. Comparability Over Time and Space

Analysis of raw data is complicated because data come from a variety of observational methods, and each observational method requires certain corrections to be made. For example, systematic biases in XBT depth measurements have recently been identified. These biases were shown to lead to erroneous estimates of ocean heat content through time. Each of the three main studies used in this indicator corrects for these XBT biases. Correction methods are slightly different among studies and are described in detail in each respective paper. More information on newly identified biases associated with XBT can be found in Barker et al. (2011).

This indicator presents three separate trend lines to compare different estimates of ocean heat content over time. Each estimate is based on analytical methods that have been applied consistently over time and space. General agreement among trend lines, despite some year-to-year variability, indicates a robust trend.

9. Sources of Uncertainty

Uncertainty measurements can be made by the organizations responsible for data collection, and they can also be made during subsequent analysis. One example of uncertainty measurements performed by an agency is available for the WOA at: www.nodc.noaa.gov/OC5/indprod.html.

Error estimates associated with each of the curves in Figure 1 are discussed in Domingues et al. (2008), Ishii and Kimoto (2009), and Levitus et al. (2009). All of the data files listed in Section 4 ("Data Availability") include a one-sigma error value for each year.

10. Sources of Variability

Weather patterns, seasonal changes, multiyear climate oscillations, and many other factors could lead to day-to-day and year-to-year variability in ocean temperature measurements at a given location. This indicator addresses some of these forms of variability by aggregating data over time and space to calculate annual values for global ocean heat content. The overall increase in ocean heat over time (as shown by all three analyses) far exceeds the range of interannual variability in ocean heat estimates.

11. Statistical/Trend Analysis

Domingues et al. (2008), Ishii and Kimoto (2009), and Levitus et al. (2009) have all calculated linear trends and corresponding error values for their respective ocean heat time series. Exact timeframes and slopes vary among the three publications, but they all reveal a generally upward trend (i.e., increasing ocean heat over time).

12. Data Limitations

Factors that may impact the confidence, application, or conclusions drawn from this indicator are as follows:

1. Data must be carefully reconstructed and filtered for biases because of different data collection techniques and uneven sampling over time and space. Various methods of correcting the data have led to slightly different versions of the ocean heat trend line.
2. In addition to differences among methods, some biases may be inherent in certain methods. The older MBT and XBT technologies have the highest uncertainty associated with measurements.
3. Limitations of data collection over time and especially over space affect the accuracy of observations. In some cases, interpolation procedures were used to complete datasets that were spatially sparse.

References

Barker, P.M., J.R. Dunn, C.M. Domingues, and S.E. Wijffels. 2011. Pressure sensor drifts in Argo and their impacts. J. Atmos. Oceanic Tech. 28:1036–1049.

Domingues, C.M., J.A. Church, N.J. White, P.J. Gleckler, S.E. Wijffels, P.M. Barker, and J.R. Dunn. 2008. Improved estimates of upper-ocean warming and multi-decadal sea-level rise. Nature 453:1090–1093.

Ishii, M., and M. Kimoto. 2009. Reevaluation of historical ocean heat content variations with time-varying XBT and MBT depth bias corrections. J. Oceanogr. 65:287–299.

Levitus, S., J.I. Antonov, T.P. Boyer, R.A. Locarnini, H.E. Garcia, and A.V. Mishonov. 2009. Global ocean heat content 1955–2008 in light of recently revealed instrumentation problems. Geophys. Res. Lett. 36:L07608.

Wijffels, S.E., J. Willis, C.M. Domingues, P. Barker, N.J. White, A. Gronell, K. Ridgway, and J.A. Church. 2008. Changing expendable bathythermograph fall rates and their impact on estimates of thermosteric sea level rise. J. Climate 21:5657–5672.

Sea Surface Temperature

Identification

1. Indicator Description

This indicator describes global trends in sea surface temperature (SST) from 1880 to 2011. Temperature is an important physical attribute of the world's oceans, with effects on global climate as well as marine ecosystems. As an example, this indicator also provides a map showing average SST across the world for the calendar year 2011.

2. Revision History

April 2010: Indicator posted
December 2011: Updated with data through 2010
January 2012: Updated with data through 2011
April 2012: Updated with revised data through 2011
July 2012: Updated example map

Data Sources

3. Data Sources

This indicator is based on the Extended Reconstructed Sea Surface Temperature (ERSST) analysis developed by the National Oceanic and Atmospheric Administration's (NOAA's) National Climatic Data Center (NCDC). The reconstruction model used here is ERSST version 3b (ERSST.v3b), which covers the years 1880 to 2011 and was described in Smith et al. (2008).

Data for the example map came from the Hadley Centre at the UK Met Office. This map is adapted from a map originally published in Sumaila et al. (2011).

4. Data Availability

NCDC and the National Center for Atmospheric Research (NCAR) provide access to monthly and annual SST and error data from the ERSST.v3b reconstruction, as well as a mapping utility that allows the user to calculate average anomalies over time and space (NOAA, 2012a). EPA used global data (all latitudes), which can be downloaded from: ftp://ftp.ncdc.noaa.gov/pub/data/cmb/mlost/pdo/. Specifically, EPA used the ASCII text file "aravg.ann.ocean.90S.90N.asc," which includes annual anomalies and error variance. A "readme" file in the same FTP directory explains how to use the ASCII file. The ERSST.v3b reconstruction is based on in situ measurements, which are available online through the International Comprehensive Ocean-Atmosphere Data Sets (ICOADS) (NOAA, 2012b).

Data for the example map were downloaded from:
http://badc.nerc.ac.uk/view/badc.nerc.ac.uk__ATOM__dataent_hadisst (UK Met Office, 2012). This website provides access to monthly global grids in ASCII format, along with documentation.

Methodology

5. Data Collection

This indicator is based on in situ instrumental measurements of water temperature worldwide from 1880 to 2011. When paired with appropriate screening criteria and bias correction algorithms, in situ records provide a reliable long-term record of temperature. The long-term sampling was not based on a scientific sampling design, but was gathered by "ships of opportunity" and other ad hoc records. Records are particularly sparse or problematic prior to the 20th century and during the two World Wars. Since about 1955, the in situ sampling has become more systematic and measurement methods have continued to improve. SST observations from drifting and moored buoys were first used in the late 1970s. Buoy observations became more plentiful following the start of the Tropical Ocean Global Atmosphere (TOGA) program in 1985. Locations have been designed to fill in data gaps where ship observations are sparse.

A summary of the relative availability, coverage, accuracy, and biases of the different measurement methods is provided in Reynolds et al. (2002). Sampling and analytical procedures are documented in several publications that can be accessed online. NOAA has documented the measurement, compilation, quality assurance, editing, and analysis for the underlying ICOADS sea surface dataset at: http://icoads.noaa.gov/publications.html.

In the original update from ERSST v2 to v3, satellite data were added to the analysis. However, ERSST version 3b no longer includes satellite data. The addition of satellite data caused problems for many users. Although the satellite data were corrected with respect to the in situ data, there was a residual cold bias that remained. The bias was strongest in the middle and high latitude Southern Hemisphere where in situ data are sparse. The residual bias led to a modest decrease in the global warming trend and modified global annual temperature rankings.

This indicator is global in scale and offers a broad overview of SST. By design, the indicator does not focus on any one region or set of sensitive areas. However, as NOAA's analysis continues to improve the resolution of data, future analyses may provide more detailed data that are more useful to the assessment of specific coastal regions and ecosystems.

Data for the example map are based on a combination of in situ instrumental measurements and remote sensing via satellite. These data are analyzed on a 1-degree global grid.

6. Indicator Derivation

This indicator is based on the ERSST, a reconstruction of historical SST using in situ data. The reconstruction methodology has undergone several stages of development and refinement. This indicator is based on the most recent data release, version 3b (ERSST.v3b).

This reconstruction involves filtering and blending datasets that use alternative measurement methods and include redundancies in space and time. Because of these redundancies, this reconstruction is able to fill spatial and temporal data gaps and correct for biases in the different measurement techniques (e.g., uninsulated canvas buckets, intakes near warm engines, uneven spatial coverage). Locations have been combined to report a single global value, based on scientifically valid techniques for averaging over areas. Specifically, data have been averaged over 5-by-5-degree grid cells as part of NOAA's Merged Land-Ocean Surface Temperature Analysis (MLOST) (www.esrl.noaa.gov/psd/data/gridded/data.mlost.html). Daily and monthly records have been averaged to find annual anomalies. Thus, the combined set of measurements is stronger than any single set. Reconstruction methods are documented in more detail by Smith et al. (2008). Smith and Reynolds (2005) discuss and analyze the similarities and differences among various reconstructions, showing that the results are generally consistent. For example, the long-term average change obtained by this method is very similar to those of the "unanalyzed" measurements and reconstructions discussed by Rayner et al. (2003).

This indicator shows the extended reconstructed data as anomalies, or differences, from a baseline "climate normal." In this case, the climate normal was defined to be the average SST from 1971 to 2000. No attempt was made to project data beyond the period during which measurements were collected.

Additional information on the compilation, data screening, reconstruction, and error analysis of the reconstructed SST data can be found at: www.ncdc.noaa.gov/ersst/.

The example map was created by obtaining monthly grids for the 12 months of 2011, then averaging the grids together to create an annual mean grid. Any grid cell that was listed as "ice-covered" for one or more months of 2011 (which meant it did not have an SST measurement for those months) was excluded from the analysis to avoid biasing the annual mean SST toward the warmer portions of the year. This step is the reason why many grid cells near the poles have been left blank.

7. Quality Assurance and Quality Control

Thorough documentation of the quality assurance and quality control (QA/QC) methods and results is available in the technical references for ERSST.v3b at NOAA's NCDC (www.ncdc.noaa.gov/ersst/).

Analysis

8. Comparability Over Time and Space

Presenting the data at a global and annual scale reduces the uncertainty and variability inherent in SST measurements, and therefore, the overall reconstruction is considered to be a good representation of global SST. This dataset covers the Earth's oceans with sufficient frequency and resolution to ensure that overall averages are not inappropriately distorted by singular events or missing data due to sparse in situ measurements or cloud cover. The confidence interval reports the estimated degree of accuracy associated with the estimates over time and suggests later measurements may be used with greater confidence than pre-20th century estimates.

Continuous improvement and greater spatial resolution can be expected in the coming years, with corresponding updates to the historical data. For example, there is a known bias during the World War II years (1941–1945), when almost all measurements were collected by U.S. Navy ships that recorded ocean intake temperatures, which can give warmer numbers than the techniques used in other years. Future efforts will aim to adjust the data more fully to account for this bias.

Comparisons by researchers Smith and Reynolds (2005) with other similar reconstructions using alternative methods yield consistent results, albeit with narrower uncertainty estimates. Hence, the indicator presented here may be more conservative than alternative methods.

9. Sources of Uncertainty

The extended reconstruction dataset includes an error variance for each year, which is associated with the biases and errors in the measurements and treatments of the data. NOAA has separated this variance into three components: high-frequency error, low-frequency error, and bias error. For this indicator, the total variance was used to calculate a 95-percent confidence interval (see Figure 1) so that the user can understand the impact of uncertainty on any conclusions that might be drawn from the time series. For each year, the square root of the error variance (the standard error) was multiplied by 1.96, and this value was added to or subtracted from the reported anomaly to define the upper and lower confidence bounds, respectively. As Figure 1 shows, the level of uncertainty has decreased dramatically in recent decades owing to better global spatial coverage and increased availability of data.

The model has largely corrected for measurement error, but some uncertainty still exists. Contributing factors include variations in sampling methodology by era as well as geographic region, and instrument error from both buoys as well as ships.

Uncertainty measurements are also available for some of the underlying data. For example, several articles have been published about uncertainties in ICOADS in situ data; these publications are available from: www.noc.soton.ac.uk/JRD/MET/coads.php.

10. Sources of Variability

Sea surface temperature varies seasonally, but this indicator has removed the seasonal signal by calculating annual averages. Temperatures can also vary as a result of interannual climate patterns such as the El Niño-Southern Oscillation.

11. Statistical/Trend Analysis

Figure 1 shows a 95 percent confidence interval that has been computed for each annual anomaly. Analysis by Smith et al. (2008) confirms that the increasing trend apparent from Figure 1 over the 20th century is statistically significant.

12. Data Limitations

Factors that may impact the confidence, application, or conclusions drawn from this indicator are as follows:

1. The 95 percent confidence interval is wider than other methods for long-term reconstructions; in mean SSTs, this interval tends to dampen anomalies.
2. The geographic resolution is coarse for ecosystem analyses but reflects long-term and global changes as well as shorter-term variability.
3. The reconstruction methods used to create this indicator remove most random "noise" in the data. However, the anomalies are also dampened when and where data are too sparse for a

reliable reconstruction. The 95 percent confidence interval reflects this "dampening effect" and uncertainty caused by possible biases in the observations.

4. Data screening results in loss of multiple observations at latitudes higher than 60 degrees north or south. Effects of screening at high latitudes are minimal in the context of the global average; the main effect is to lessen anomalies and widen confidence intervals.

References

NOAA (National Oceanic and Atmospheric Administration). 2012a. Extended reconstructed sea surface temperature (ERSST.v3b). National Climatic Data Center. Accessed April 2012. www.ncdc.noaa.gov/ersst/.

NOAA (National Oceanic and Atmospheric Administration). 2012b. International comprehensive ocean-atmosphere data sets (ICOADS). Accessed April 2012. http://icoads.noaa.gov/.

Rayner, N.A., D.E. Parker, E.B. Horton, C.K. Folland, L.V. Alexander, D.P. Rowell, E.C. Kent, and A. Kaplan. 2003. Global analyses of sea surface temperature, sea ice, and night marine air temperature since the late nineteenth century. J. Geophys. Res. 108:4407.

Smith, T.M., and R.W. Reynolds. 2005. A global merged land air and sea surface temperature reconstruction based on historical observations (1880–1997). J. Climate 18(12):2021-2036. www.ncdc.noaa.gov/oa/climate/research/Smith-Reynolds-dataset-2005.pdf.

Smith, T.M., R.W. Reynolds, T.C. Peterson, and J. Lawrimore. 2008. Improvements to NOAA's historical merged land-ocean surface temperature analysis (1880–2006). J. Climate 21(10):2283–2296. www.ncdc.noaa.gov/ersst/papers/SEA.temps08.pdf.

Sumaila, U.R., W.W.L. Cheung, V.W.Y. Lam, D. Pauly, and S. Herrick. 2011. Climate change impacts on the biophysics and economics of world fisheries. Nature Climate Change 1:449–456.

UK Met Office. 2012. Hadley Centre, HadISST 1.1: Global sea ice coverage and sea surface temperature (1870-present). NCAS British Atmospheric Data Centre. Accessed May 2012. http://badc.nerc.ac.uk/view/badc.nerc.ac.uk__ATOM__dataent_hadisst.

Xue, Y., T.M. Smith, and R.W. Reynolds. 2003. Interdecadal changes of 30-yr SST normals during 1871–2000. J. Climate 16:1601-1612. www.ncdc.noaa.gov/ersst/papers/xue-etal.pdf.

Sea Level

Identification

1. Indicator Description

This indicator describes how sea level has changed since 1880. Rising sea levels are associated with climate change, and they can affect human activities in coastal areas and can alter ecosystems.

Components of this indicator include:

- Average absolute sea level change of the world's oceans since 1880 (Figure 1)
- Trends in relative sea level change along U.S. coasts over the past half-century (Figure 2)

2. Revision History

April 2010: Indicator posted
December 2011: Updated with data through 2009
May 2012: Updated with altimeter data through 2011 from a new source and tide gauge data from 1960 to 2011
June 2012: Updated with long-term reconstruction data through 2011

Data Sources

3. Data Sources

Figure 1. Global Average Absolute Sea Level Change, 1880–2011

Figure 1 presents a reconstruction of absolute sea level developed by Australia's Commonwealth Scientific and Industrial Research Organisation (CSIRO). This reconstruction is based on two main data sources:

- Satellite data from the TOPography EXperiment (TOPEX)/Poseidon, Jason-1, and Jason-2 satellite altimeters, operated by the National Aeronautics and Space Administration (NASA) and France's Centre National d'Etudes Spatiales (CNES).
- Tide gauge measurements compiled by the Permanent Service for Mean Sea Level (PSMSL), which includes over a century's worth of daily and monthly tide gauge data.

Figure 1 also presents the National Oceanic and Atmospheric Administration's (NOAA's) analysis of altimeter data from the TOPEX/Poseidon, Jason-1 and -2, GEOSAT Follow-On (GFO), Envisat, and European Remote Sensing (ERS) 2 satellite missions.

Figure 2. Relative Sea Level Change Along U.S. Coasts, 1960–2011

Figure 2 presents tide gauge trends calculated by NOAA. The original data come from the National Water Level Observation Network (NWLON), operated by the Center for Operational Oceanographic Products and Services (CO-OPS) within NOAA's National Ocean Service (NOS).

4. Data Availability

Figure 1. Global Average Absolute Sea Level Change, 1880–2011

The CSIRO long-term tide gauge reconstruction has been published online in graph form at: www.cmar.csiro.au/sealevel, and the data are posted on CSIRO's website at: www.cmar.csiro.au/sealevel/sl_data_cmar.html. The same results were also published in Church and White (2011). CSIRO's website also provides a list of tide gauges that were used to develop the long-term tide gauge reconstruction.

At the time this indicator was published, CSIRO's website presented data through 2009. EPA obtained an updated version of the analysis with data through 2011 from Dr. Neil White at CSIRO.

The satellite time series was obtained from NOAA's Laboratory for Satellite Altimetry, which maintains an online repository of sea level data (NOAA, 2012). The data file for this indicator was downloaded from: http://ibis.grdl.noaa.gov/SAT/SeaLevelRise/slr/slr_sla_gbl_free_all_66.csv. Underlying satellite measurements can be obtained from NASA's online database (NASA, 2012). The reconstructed tide gauge time series is based on data from the PSMSL database, which can be accessed online at: www.pol.ac.uk/psmsl.

Figure 2. Relative Sea Level Change Along U.S. Coasts, 1960–2011

The relative sea level map is based on individual station measurements that can be accessed through NOAA's "Sea Levels Online" website at: http://tidesandcurrents.noaa.gov/sltrends/sltrends.shtml. This website also presents an interactive map that illustrates sea level trends over different timeframes. NOAA has not yet published the table of 1960–2011 trends that it provided to EPA for this indicator; however, a user could reproduce these numbers from the publicly available data cited above. NOAA published an earlier version of this trend analysis in a technical report on sea level variations of the United States from 1854 to 1999 (NOAA, 2001). EPA obtained the updated 1960–2011 analysis from the lead author of NOAA (2001), Chris Zervas.

Methodology

5. Data Collection

This indicator presents absolute and relative sea level changes. Absolute sea level change (Figure 1) represents only the sea height, whereas relative sea level change (Figure 2) is defined as the change in sea height relative to land. Land surfaces move up or down in many locations around the world due to natural geologic processes (such as uplift and subsidence) and human activities that can cause ground to sink (e.g., from extraction of groundwater or hydrocarbons that supported the surface).

Sea level has traditionally been measured using tide gauges, which are mechanical measuring devices located along the shore. These devices measure the change in sea level relative to the land surface, which means the resulting data reflect both the change in absolute sea surface height and the change in local land levels. Satellite measurement of land and sea surface heights (altimetry) began several decades ago; this technology allows for measurement of changes in absolute sea level. Tide gauge data can be converted to absolute change (as in Figure 1) through a series of adjustments as described in Section 6.

The two types of sea level data (relative and absolute) complement each other, and each is useful for different purposes. Relative sea level trends show how sea level change and vertical land movement together are likely to affect coastal lands and infrastructure, while absolute sea level trends provide a more comprehensive picture of the volume of water in the world's oceans, how it is changing, and how these changes relate to other observed or predicted changes in global systems (e.g., increasing ocean heat content and melting polar ice caps). Tide gauges provide more precise local measurements, while satellite data provide more complete spatial coverage. Tide gauges are used to help calibrate satellite data. For more discussion of the advantages and limitations of each type of measurement, see Cazenave and Nerem (2004).

Tide Gauge Data

Tide gauge sampling takes place at sub-daily resolution at sites around the world. Some locations have had continuous tide gauge measurements since the 1800s.

Tide gauge data for Figure 1 were collected by numerous networks of tide gauges around the world. The number of stations included in the analysis varies from year to year, ranging from fewer than 20 locations in the 1880s to more than 200 locations during the 1980s. Pre-1880 data were not included in the reconstruction because of insufficient tide gauge coverage. These measurements are documented by the PSMSL, which compiled data from various networks. The PSMSL data catalogue provides documentation for these measurements at: www.pol.ac.uk/psmsl/datainfo.

Tide gauge data for Figure 2 come from NOAA's NWLON. The NWLON is composed of 175 long-term, continuously operating tide gauge stations located along the United States coast, including the Great Lakes and islands in the Atlantic and Pacific Oceans. The map in Figure 2 shows trends for 68 stations along the ocean coasts that had sufficient data over the period from 1960 to 2011. NOAA (2001) describes these data and how they were collected. Data collection methods are documented in a series of manuals and standards that can be accessed at: www.co-ops.nos.noaa.gov/pub.html#sltrends.

Satellite Data

Satellite altimetry has revealed that the rate of change in absolute sea level differs around the globe (Cazenave and Nerem, 2004). Factors that lead to changes in sea level include astronomical tides; variations in atmospheric pressure, wind, river discharge, ocean circulation, and water density (associated with temperature and salinity); and added or extracted water volume due to the melting of ice or changes in the storage of water on land in reservoirs and aquifers.

Data for this indicator came from the following satellite missions:

- TOPEX/Poseidon began collecting data in late 1992; Jason replaced TOPEX/Poseidon around 2002. For more information about the TOPEX/Poseidon and Jason missions, see NASA's website at: http://sealevel.jpl.nasa.gov/missions/.

- The U.S. Navy launched GFO in 1998, and altimeter data are available from 2000 through 2006. For more information about the GFO missions, see NASA's website at: http://gcmd.nasa.gov/records/GCMD_GEOSAT_FOLLOWON.html.

- The European Space Agency (ESA) launched ERS-2 in 1995, and its sea level data are available from 1995 through 2003. More information about the mission can be found on ESA's website at: https://earth.esa.int/web/guest/missions/esa-operational-eo-missions/ers.

- ESA launched Envisat in 2002, and this indicator includes data from 2002 through 2010. More information about Envisat can be found on ESA's website at: https://earth.esa.int/web/guest/missions/esa-operational-eo-missions/envisat.

TOPEX/Poseidon and Jason satellite altimeters each cover the entire globe between 66 degrees south and 66 degrees north with 10-day resolution. Some of the other satellites have different resolutions and orbits. For example, Envisat is a polar-orbiting satellite.

6. Indicator Derivation

Satellite Data for Figure 1. Global Average Absolute Sea Level Change, 1880–2011

NOAA processed all of the satellite measurements so they could be combined into a single time series. In doing so, NOAA limited its analysis to data between 66 degrees south and 66 degrees north, which is the area with the most complete satellite coverage.

Researchers removed spurious data points. They also estimated and removed inter-satellite biases to allow for a continuous time series over the time of transition from TOPEX/Poseidon to Jason-1 and -2. A discussion of the methods for calibrating satellite data is available in Leuliette et al. (2004) for TOPEX/Poseidon data and in Chambers et al. (2003) for Jason data. Also see Nerem et al. (2010).

Data were adjusted using an inverted barometer correction, which corrects for air pressure differences, along with an algorithm to remove average seasonal signals. These corrections reflect standard procedures for analyzing sea level data and are documented in the metadata for the dataset. The data were not corrected for glacial isostatic adjustment (GIA)—an additional factor explained in more detail below.

NOAA provided individual measurements, spaced approximately 10 days apart (or more frequent, depending on how many satellite missions were collecting data during the same timeframe). EPA generated monthly averages based on all available data points, then combined these monthly averages to determine annual averages. EPA chose to calculate annual averages from monthly averages in order to reduce the potential for biasing the annual average toward a portion of the year in which measurements were spaced more closely together (e.g., due to the launch of an additional satellite mission).

The analysis of satellite data has improved over time, which has led to a high level of confidence in the associated measurements of sea level change. Further discussion can be found in Cazenave and Nerem (2004), Miller and Douglas (2004), and Church and White (2011).

Several other groups have developed their own independent analyses of satellite altimeter data. Although all of these interpretations have appeared in the literature, EPA has chosen to include only one (NOAA) in the interest of keeping this indicator straightforward and accessible to readers. Other organizations that publish altimeter-based data include:

- The University of Colorado at Boulder: http://sealevel.colorado.edu/
- AVISO (France): www.aviso.oceanobs.com/en/news/ocean-indicators/mean-sea-level/
- CSIRO: www.cmar.csiro.au/sealevel/

Tide Gauge Reconstruction for Figure 1. Global Average Absolute Sea Level Change, 1880–2011

CSIRO developed the long-term tide gauge reconstruction using a series of adjustments to convert relative tide gauge measurements into an absolute global mean sea level trend. Church and White (2011) describe the methods used, which include data screening; calibration with satellite altimeter data to establish patterns of spatial variability; and a correction for GIA, which represents the ongoing change in the size and shape of the ocean basins associated with changes in surface loading. On average, the world's ocean crust is sinking in response to the transfer of mass from the land to the ocean following the retreat of the continental ice sheets after the Last Glacial Maximum (approximately 20,000 years ago). Worldwide, on average, the ocean crust is sinking at a rate of approximately 0.3 mm per year. By correcting for GIA, the resulting curve actually reflects the extent to which sea level *would* be rising if the ocean basins were not becoming larger (deeper) at the same time. For more information about GIA and the value of correcting for it, see: http://sealevel.colorado.edu/content/what-glacial-isostatic-adjustment-gia-and-why-do-you-correct-it.

Seasonal signals have been removed, but no inverse barometer (air pressure) correction has been applied because a suitable long-term global air pressure dataset is not available. Figure 1 shows annual average change in the form of an anomaly. EPA has labeled the graph as "cumulative sea level change" for the sake of clarity.

The tide gauge reconstruction required the use of a modeling approach to derive a global average from individual station measurements. This approach allowed the authors to incorporate data from a time-varying array of tide gauges in a consistent way.

Figure 2. Relative Sea Level Change Along U.S. Coasts, 1960–2011

Figure 2 shows relative sea level change for 68 tide gauges with adequate data for the period from 1960 to 2011. Sites were selected if they began recording data in 1960 or earlier and if data were available through 2011. Sites in south-central Alaska between Kodiak Island and Yakutat were excluded from the analysis because they have exhibited nonlinear behavior since a major earthquake occurred in 1964.

Extensive discussion of this network and the tide gauge data analysis can be found in NOAA (2001) and additional sources available from the CO-OPS website at: http://tidesandcurrents.noaa.gov. The process of generating Figure 2 involved only simple mathematics. NOAA used monthly sea level means to calculate a long-term annual rate of change for each station, which was determined by linear regression.

EPA multiplied the annual rate of change by the length of the analysis period (52 years) to determine total change.

7. Quality Assurance and Quality Control

Satellite data processing involves extensive quality assurance and quality control (QA/QC) protocols—for example, to check for instrumental drift by comparing with tide gauge data (note that no instrumental drift has been detected for many years). The papers cited in Sections 5 and 6 document all such QA/QC procedures.

Church and White (2011) and earlier publications cited therein describe steps that were taken to select the highest-quality sites and correct for various sources of potential error in tide gauge measurements used for the long-term reconstruction in Figure 1. QA/QC procedures for the U.S. tide gauge data in Figure 2 are described in various publications available at: www.co-ops.nos.noaa.gov/pub.html#sltrends.

Analysis

8. Comparability Over Time and Space

Figure 1. Global Average Absolute Sea Level Change, 1880–2011

Satellite data were collected by several different satellite altimeters over different timespans. Steps have been taken to calibrate the results and remove biases over time, and NOAA made sure to restrict its analysis to the portion of the globe between 66 degrees south and 66 degrees north, where coverage is most complete.

The number of tide gauges collecting data has changed over time. The methods used to reconstruct a long-term trend, however, adjust for these changes.

The most notable difference between the two time series displayed in Figure 1 is that the long-term reconstruction includes a GIA correction, while the altimeter series does not. The uncorrected (altimeter) time series gives the truest depiction of how the surface of the ocean is changing in relation to the center of the Earth, while the corrected (long-term) time series technically shows how the volume of water in the ocean is changing. A very small portion of this volume change is not observed as absolute sea level rise (although most is) because of the GIA correction. Some degree of GIA correction is needed for a tide-gauge-based reconstruction in order to adjust for the effects of vertical crust motion.

Figure 2. Relative Sea Level Change Along U.S. Coasts, 1960–2011

Only the 68 stations with sufficient data between 1960 and 2011 were used to show sea level trends. However, tide gauge measurements at specific locations are not indicative of broader changes over space, and the network is not designed to achieve uniform spatial coverage. Rather, the gauges tend to be located at major port areas along the coast, and measurements tend to be more clustered in heavily populated areas like the Mid-Atlantic coast. Nevertheless, in many areas it is possible to see consistent patterns across numerous gauging locations—for example, rising relative sea level all along the U.S. Atlantic and Gulf Coasts.

9. Sources of Uncertainty

Figure 1. Global Average Absolute Sea Level Change, 1880–2011

Figure 1 shows bounds of +/- one standard deviation around the long-term tide gauge reconstruction. For more information about error estimates related to the tide gauge reconstruction, see Church and White (2011).

Leuliette et al. (2004) provide a general discussion of uncertainty for satellite altimeter data. The Jason instrument currently provides an estimate of global mean sea level every 10 days, with an uncertainty of 3 to 4 millimeters.

Figure 2. Relative Sea Level Change Along U.S. Coasts, 1960–2011

Standard deviations for each station-level trend estimate were included in the dataset provided to EPA by NOAA. Overall, with approximately 50 years of data, the 95 percent confidence interval around the long-term rate of change at each station is approximately +/- 0.5 mm per year. Error measurements for each tide gauge station are also described in NOAA (2001), but many of the estimates in that publication pertain to longer-term time series (i.e., the entire period of record at each station, not the 52-year period covered by this indicator).

General Discussion

Uncertainties in the data do not impact the overall conclusions. Tide gauge data do present challenges, as described by Parker (1992) and various publications available from: www.co-ops.nos.noaa.gov/pub.html#sltrends. Since 2001, there has been some disagreement and debate over the reliability of the tide gauge data and estimates of global sea level rise trends from these data (Cabanes et al., 2001). However, further research on comparisons of satellite data with tide gauge measurements and on improved estimates of contributions to sea level rise by sources other than thermal expansion—and by Alaskan glaciers in particular—have largely resolved the question (Cazenave and Nerem, 2004; Miller and Douglas, 2004). These studies have in large part closed the gap between different methods of measuring sea level change, although further improvements are expected as more measurements and longer time series become available.

10. Sources of Variability

Changes in sea level can be influenced by multi-year cycles such as El Niño/La Niña and the Pacific Decadal Oscillation, which affect coastal ocean temperatures, salt content, winds, atmospheric pressure, and currents. The satellite data record is of insufficient length to distinguish medium-term variability from long-term change, which is why the satellite record in Figure 1 has been supplemented with a longer-term reconstruction based on tide gauge measurements.

11. Statistical/Trend Analysis

Figure 1. Global Average Absolute Sea Level Change, 1880–2011

The indicator text refers to long-term rates of change, which were calculated using ordinary least-squares regression, a commonly used method of trend analysis. The long-term tide gauge reconstruction trend reflects an average increase of 0.07 inches per year. The 1993–2011 trend is 0.13 inches per year for the reconstruction, and the 1993–2011 trend for the NOAA altimeter-based time series is 0.11 inches per year. Church and White (2011) provide more information about long-term rates of change and their confidence bounds.

Figure 2. Relative Sea Level Change Along U.S. Coasts, 1960–2011

U.S. relative sea level results have been generalized over time by calculating long-term rates of change for each station using ordinary least-squares regression. No attempt was made to interpolate these data geographically.

12. Data Limitations

Factors that may impact the confidence, application, or conclusions drawn from this indicator are as follows:

1. Relative sea level trends represent a combination of absolute sea level change and local changes in land elevation. Tide gauge measurements such as those presented in Figure 2 generally cannot distinguish between these two influences without an accurate measurement of vertical land motion nearby.
2. Some changes in relative and absolute sea level may be due to multiyear cycles such as El Niño/La Niña and the Pacific Decadal Oscillation, which affect coastal ocean temperatures, salt content, winds, atmospheric pressure, and currents. The satellite data record is of insufficient length to distinguish medium-term variability from long-term change, which is why the satellite record in Figure 1 has been supplemented with a longer-term reconstruction based on tide gauge measurements.
3. Satellite data do not provide sufficient spatial resolution to resolve sea level trends for small water bodies, such as many estuaries, or for localized interests, such as a particular harbor or beach.
4. Most satellite altimeter tracks span the area from 66 degrees north latitude to 66 degrees south, so they cover about 90 percent of the ocean surface, not the entire ocean.

References

Cabanes, C., A. Cazenave, and C. Le Provost. 2001. Sea level rise during past 40 years determined from satellite and in situ observations. Science 294(5543):840–842.

Cazenave, A., and R.S. Nerem. 2004. Present-day sea level change: Observations and causes. Rev. Geophys. 42(3):1–20.

Chambers, D.P., S.A. Hayes, J.C. Ries, and T.J. Urban. 2003. New TOPEX sea state bias models and their effect on global mean sea level. J. Geophys. Res. 108(C10):3-1–3-7.

Church, J.A., and N.J. White. 2011. Sea-level rise from the late 19th to the early 21st century. Surv. Geophys. 32:585–602.

Leuliette, E., R. Nerem, and G. Mitchum. 2004. Calibration of TOPEX/Poseidon and Jason altimeter data to construct a continuous record of mean sea level change. Marine Geodesy 27(1–2):79–94.

Miller, L., and B.C. Douglas. 2004. Mass and volume contributions to twentieth-century global sea level rise. Nature 428(6981):406–409. www.grdl.noaa.gov/SAT/pubs/papers/2004nature.pdf.

NASA (National Aeronautics and Space Administration). 2012. Ocean surface topography from space. Accessed May 2012. http://sealevel.jpl.nasa.gov/.

Nerem, R.S., D.P. Chambers, C. Choe, and G.T. Mitchum. 2010. Estimating mean sea level change from the TOPEX and Jason altimeter missions. Marine Geodesy 33:435–446.

NOAA (National Oceanic and Atmospheric Administration). 2001. Sea level variations of the United States 1854–1999. NOAA Technical Report NOS CO-OPS 36. http://tidesandcurrents.noaa.gov/publications/techrpt36.pdf.

NOAA (National Oceanic and Atmospheric Administration). 2012. Laboratory for Satellite Altimetry: Sea level rise. Accessed May 2012. http://ibis.grdl.noaa.gov/SAT/SeaLevelRise/LSA_SLR_timeseries_global.php.

Parker, B.B. 1992. Sea level as an indicator of climate and global change. Mar. Technol. Soc. J. 25(4):13–24.

Ocean Acidity

Identification

1. Indicator Description

This indicator shows recent trends in acidity levels in the ocean at three key locations. The indicator also presents changes in aragonite saturation by comparing historical data with the most recent decade. Ocean acidity and aragonite saturation levels are strongly affected by the amount of carbon dissolved in the water, which is directly related to the amount of carbon dioxide in the atmosphere.

Components of this indicator include:

- Recent trends in ocean carbon dioxide and acidity levels (Figure 1)
- Historical changes in the aragonite saturation of the world's oceans (Figure 2)

2. Revision History

April 2010: Indicator posted
May 2012: Figure 1 data updated; new Figure 2 source and metric

Data Sources

3. Data Sources

Figure 1 includes trend lines from three different ocean time series: the Bermuda Atlantic Time-Series Study (BATS); the European Station for Time-Series in the Ocean, Canary Islands (ESTOC); and the Hawaii Ocean Time-Series (HOT).

Figure 2 contains aragonite saturation (Ω_{ar}) calculations derived from atmospheric carbon dioxide records from ice cores and observed atmospheric concentrations at Mauna Loa, Hawaii. These atmospheric carbon dioxide measurements are processed using the Community Earth Systems Model (CESM), maintained by the National Center for Atmospheric Research (NCAR), to determine historical concentrations of carbonate (CO_3^{2-}) in seawater. Along with salinity and temperature, this value can be used to calculate historical Ω_{ar} values.

4. Data Availability

Figure 1 compiles pCO_2 (the mean seawater CO_2 partial pressure in µatm) and pH data from three sampling programs in the Atlantic and Pacific Oceans. Raw data from the three ocean sampling programs are publicly available online. In the case of Bermuda and the Canary Islands, updated data were procured directly from the scientists leading those programs. BATS data and descriptions are available at: http://bats.bios.edu/bats_form_bottle.html. ESTOC data can be downloaded from: www.eurosites.info/estoc/data.php. HOT data were downloaded from the HOT Data Organization and

Graphical System website at: http://hahana.soest.hawaii.edu/hot/products/products.html. Additionally, annual HOT data reports are available at: http://hahana.soest.hawaii.edu/hot/reports/reports.html.

The map in Figure 2 is derived from the same source data as NOAA's Ocean Acidification "Science on a Sphere" video simulation at: http://sos.noaa.gov/Datasets/list.php?category=Ocean (Feely et al., 2009). EPA obtained the map data from Dr. Sarah Cooley of the Woods Hole Oceanographic Institution (WHOI).

Methodology

5. Data Collection

Figure 1. Ocean Carbon Dioxide Levels and Acidity, 1983–2011

This indicator reports on the pH of the upper 5 meters of the ocean and the corresponding partial pressure of dissolved carbon dioxide (pCO_2). Each dataset covers a different time period:

- BATS data used in this indicator are available from 1983 to 2011. Samples were collected from two locations in the Atlantic Ocean near Bermuda (BATS and Hydrostation S), located at (31°43' N, 64°10' W) and (32°10' N, 64°30' W), respectively (see: http://bats.bios.edu/bats_location.html).
- ESTOC data are available from 1995 to 2009. ESTOC is located at (29°10' N, 15°30' W) in the Atlantic Ocean.
- HOT data are available from 1988 to 2010. The HOT station is located at (23° N, 158° W) in the Pacific Ocean.

At the BATS and HOT stations, dissolved inorganic carbon (DIC) and total alkalinity (TA) were measured directly from water samples. DIC accounts for the carbonate and bicarbonate ions that occur when CO_2 dissolves to form carbonic acid, while total alkalinity measures the buffering capacity of the water, which is affected by the addition of a weak acid such as carbonic acid. At ESTOC, pH and alkalinity were measured directly (Bindoff et al., 2007).

Each station followed internally consistent sampling protocols over time. Bates et al. (2012) describe the sampling plan for BATS. Further information on BATS sampling methods is available at: http://bats.bios.edu. ESTOC sampling procedures are described by González-Dávila et al. (2010). HOT sampling procedures are described in documentation available at: http://hahana.soest.hawaii.edu/hot/hot_jgofs.html and: http://hahana.soest.hawaii.edu/hot/products/HOT_surface_CO2_readme.pdf.

Figure 2. Changes in Aragonite Saturation of the World's Oceans, 1880–2012

The map in Figure 2 shows the estimated change in sea surface Ω_{ar} from 1880 to 2012. Aragonite saturation values are calculated in a multi-step process that originates from historical atmospheric CO_2 concentrations that are built into the model (the CESM). As documented in Orr et al. (2001), this model uses historical atmospheric CO_2 concentrations based on ice cores and atmospheric measurements (collected at Mauna Loa, Hawaii).

6. Indicator Derivation

Figure 1. Ocean Carbon Dioxide Levels and Acidity, 1983–2011

At BATS and HOT stations, pH and pCO_2 values were calculated based on DIC and TA measurements from water samples. BATS analytical procedures are described by Bates et al. (2012). HOT analytical procedures are described in documentation available at: http://hahana.soest.hawaii.edu/hot/hot_jgofs.html and: http://hahana.soest.hawaii.edu/hot/products/HOT_surface_CO2_readme.pdf. At ESTOC, pCO_2 was calculated from direct measurements of pH and alkalinity. ESTOC analytical procedures are described by González-Dávila et al. (2010). For all three locations, Figure 1 shows in situ measured or calculated values for pCO_2 and pH, as opposed to values adjusted to a standard temperature.

The lines in Figure 1 connect points that represent individual sampling events. No attempt was made to generalize data spatially or portray data beyond the temporal windows in which measurements were made. Unlike some figures in the published source studies, the data shown in Figure 1 are not adjusted for seasonal variability. The temporal interval between sampling events is somewhat irregular at all three locations, so moving averages and monthly or annual averages based on these data could be misleading. Thus, EPA elected to show individual measurements in Figure 1.

Figure 2. Changes in Aragonite Saturation of the World's Oceans, 1880–2012

The map in Figure 2 was created by Dr. Sarah Cooley (Woods Hole Oceanographic Institution) using the CESM, which is available publicly at: www.cesm.ucar.edu/models/. Atmospheric CO_2 concentrations were processed by the CESM to calculate historical ocean attributes, including monthly salinity, temperature, and CO_3^{2-} concentrations, on an approximately 1° by 1° grid. Next, these monthly model outputs were used to approximate concentrations of the calcium ion (Ca^{2+}) as a function of salt (Millero, 1982) and calculate aragonite solubility according to Mucci (1983). The resulting aragonite saturation state was calculated using a standard polynomial solver for MATLAB, which was developed by Dr. Richard Zeebe of the University of Hawaii. This solver is available at: www.soest.hawaii.edu/oceanography/faculty/zeebe_files/CO2_System_in_Seawater/csys.html.

Aragonite saturation state is represented as Ω_{ar}, which is defined as:

$$\Omega_{ar} = [Ca^{2+}][CO_3^{2-}] / K'_{sp}$$

The numerator represents the product of the observed concentrations of calcium and carbonate ions. K'_{sp} is the apparent solubility product, which is a constant that is equal to $[Ca^{2+}][CO_3^{2-}]$ at equilibrium for a given set of temperature, pressure, and salinity conditions. Thus, Ω_{ar} is a unitless ratio that compares the observed concentrations of calcium and carbonate ions dissolved in the water with the concentrations that would be observed under fully saturated conditions. An Ω_{ar} value of 1 represents full saturation, while a value of 0 indicates that no aragonite is dissolved in the water. Ocean water at the surface can be supersaturated with aragonite, however, so it is possible to have an Ω_{ar} value greater than 1, and it is also possible to experience a decrease over time yet still have water that is supersaturated.

For Figure 2, monthly model outputs were averaged by decade before calculating Ω_{ar} for each grid cell. The resulting map is based on averages for two decades: 1880–1889 (a baseline) and 2003–2012 (the most recent decade available). Figure 2 shows the change in Ω_{ar} between the earliest (baseline) decade and the most recent decade. It is essentially an endpoint-to-endpoint comparison, but using decadal averages instead of individual years offers some protection against inherent year-to-year variability. The map has approximately 1° by 1° resolution.

7. Quality Assurance and Quality Control

Quality assurance and quality control (QA/QC) steps are followed during data collection and data analysis. These procedures are described in the documentation listed in Sections 5 and 6.

Analysis

8. Comparability Over Time and Space

Figure 1. Ocean Carbon Dioxide Levels and Acidity, 1983–2011

BATS, ESTOC, and HOT each use different methods to determine pH and pCO_2, though each individual sampling program uses well-established methods that are consistent over time.

Figure 2. Changes in Aragonite Saturation of the World's Oceans, 1880–2012

The CESM calculates data for all points in the Earth's oceans using comparable methods. Atmospheric CO_2 concentration values differ in origin depending on their age (i.e., older values from ice cores and more recent values from direct atmospheric measurement). However, all biogeochemical calculations performed by the CESM use the atmospheric CO_2 values in the same manner.

9. Sources of Uncertainty

Figure 1. Ocean Carbon Dioxide Levels and Acidity, 1983–2011

Uncertainty measurements can be made for raw data as well as analyzed trends. Details on uncertainty measurements can be found in the following documents and references therein: Bindoff et al. (2007), Bates et al. (2012), Dore et al. (2009), and González-Dávila et al. (2010).

Figure 2. Changes in Aragonite Saturation of the World's Oceans, 1880–2012

Uncertainty and confidence for CESM calculations, as they compare with real-world observations, are measured and analyzed in Doney et al. (2009). Uncertainty for the approximation of Ca^{2+} and aragonite solubility are documented in Millero (1982) and Mucci (1983), respectively.

10. Sources of Variability

Aragonite saturation, pH, and pCO_2 are properties of seawater that vary with temperature and salinity. Therefore, these parameters naturally vary over space and time. Variability in ocean surface pH and pCO_2 data has been associated with regional changes in the natural carbon cycle influenced by changes in ocean circulation, climate variability (seasonal changes), and biological activity (Bindoff et al., 2007).

Figure 1. Ocean Carbon Dioxide Levels and Acidity, 1983–2011

Variability associated with seasonal signals is still present in the data presented in Figure 1. This seasonal variability can be identified by the oscillating line that connects sampling events for each site.

Figure 2. Changes in Aragonite Saturation of the World's Oceans, 1880–2012

Figure 2 shows how changes in Ω_{ar} vary geographically. Monthly and yearly variations in CO_2 concentrations, temperature, salinity, and other relevant parameters have been addressed by calculating decadal averages.

11. Statistical/Trend Analysis

This indicator does not report on the slope of the apparent trends in ocean acidity and pCO_2 in Figure 1. The long-term trends in Figure 2 are based on an endpoint-to-endpoint comparison between the first decade of widespread data (the 1880s) and the most recent 10 years of data (2003–2012). The statistical significance of these trends has not been calculated.

12. Data Limitations

Factors that may impact the confidence, application, or conclusions drawn from this indicator are as follows:

1. Carbon variability exists in the surface layers of the ocean as a result of changing surface temperatures, mixing of layers as a result of ocean circulation, and other seasonal variations.
2. Changes in ocean pH caused by the uptake of atmospheric carbon dioxide tend to occur slowly relative to natural fluctuations, so the full effect of atmospheric carbon dioxide concentrations on ocean pH may not be seen for many decades, if not centuries.
3. Ocean chemistry is not uniform throughout the world's oceans, so local conditions could cause a pH measurement to seem incorrect or abnormal in the context of the global data.
4. Although closely tied to atmospheric concentrations of CO_2, aragonite saturation is not exclusively controlled by atmospheric CO_2, as salinity and temperature are also factored into the calculation.

References

Bates, N.R., M.H.P. Best, K. Neely, R. Garley, A.G. Dickson, and R.J. Johnson. 2012. Detecting anthropogenic carbon dioxide uptake and ocean acidification in the North Atlantic Ocean. Biogeosciences Discuss. 9:989–1019.

Bindoff, N.L., J. Willebrand, V. Artale, A, Cazenave, J. Gregory, S. Gulev, K. Hanawa, C. Le Quéré, S. Levitus, Y. Nojiri, C.K. Shum, L.D. Talley, and A. Unnikrishnan. 2007. Observations: Oceanic climate change and sea level. In: Climate change 2007: The physical science basis (fourth assessment report). Cambridge, United Kingdom: Cambridge University Press.

Doney, S.C., I. Lima, J.K. Moore, K. Lindsay, M.J. Behrenfeld, T.K. Westberry, N. Mahowald, D.M. Glober, and T. Takahashi. 2009. Skill metrics for confronting global upper ocean ecosystem-biogeochemistry models against field and remote sensing data. J. Mar. Syst. 76(1–2):95–112.

Dore, J.E., R. Lukas, D.W. Sadler, M.J. Church, and D.M. Karl. 2009. Physical and biogeochemical modulation of ocean acidification in the central North Pacific. Proc. Natl. Acad. Sci. USA 106:12235–12240.

Feely, R.A., S.C. Doney, and S.R. Cooley. 2009. Ocean acidification: Present conditions and future changes in a high-CO_2 world. Oceanography 22(4):36–47.

González-Dávila, M., J.M. Santana-Casiano, M.J. Rueda, and O. Llinás. 2010. The water column distribution of carbonate system variables at the ESTOC site from 1995 to 2004. Biogeosciences Discuss. 7:1995–2032.

Millero, F.J. 1982. The thermodynamics of seawater. Part I: The PVT properties. Ocean Science and Engineering 7(4):403–460.

Mucci, A. 1983. The solubility of calcite and aragonite in seawater at various salinities, temperatures, and one atmosphere total pressure. Amer. Jour. Sci. 283:780–799.

Orr, J.C., E. Maier-Reimer, U. Mikolajewicz, P. Monfray, J.L. Sarmiento, J.R. Toggweiler, N.K. Taylor, J. Palmer, N. Gruber, C.L. Sabine, C.L. Le Quéré, R.M. Key, and J. Boutin. 2001. Estimates of anthropogenic carbon uptake from four three-dimensional global ocean models. Global Biogeochem. Cycles 15(1):43–60.

Sabine, C.L., R.A. Feely, N. Gruber, R.M. Key, K. Lee, J.L. Bullister, R. Wanninkhof, C.S. Wong, D.W.R. Wallace, B. Tilbrook, F.J. Millero, T.-H. Peng, A. Kozyr, T. Ono, and A.F. Rios. 2004. The oceanic sink for anthropogenic CO_2. Science 305(5682):367–371.

Sabine, C.L., R.M. Key, A. Kozyr, R.A. Feely, R. Wanninkhof, F.J. Millero, T.-H. Peng, J.L. Bullister, and K. Lee. 2005. Global Ocean Data Analysis Project: Results and data. ORNL/CDIAC-145, NDP-083. Carbon Dioxide Information Analysis Center, Oak Ridge National Laboratory, U.S. Department of Energy.

Arctic Sea Ice

Identification

1. Indicator Description

This indicator tracks the extent and age of sea ice in the Arctic Ocean. The extent of area covered by Arctic sea ice is considered a particularly sensitive indicator of global climate because a warmer climate will reduce the amount of sea ice present. The proportion of sea ice in each age category can indicate the relative stability of Arctic conditions as well as susceptibility to melting events.

Components of this indicator include:

- Changes in the September average extent of sea ice in the Arctic Ocean since 1979 (Figure 1)
- Changes in the proportion of Arctic sea ice in various age categories at the September weekly minimum since 1983 (Figure 2)

2. Revision History

April 2010: Indicator of Arctic sea ice extent posted
December 2011: Updated with data through 2011; added age of ice
October 2012: Updated with data through 2012

Data Sources

3. Data Sources

Figure 1 (extent of sea ice) is based on monthly average sea ice extent data provided by NSIDC. NSIDC's data are derived from satellite imagery collected and processed by the National Aeronautics and Space Administration (NASA). NSIDC also provided Figure 2 data (age distribution of sea ice), which are derived from weekly NASA satellite imagery and processed by the team of Maslanik and Tschudi at the University of Colorado–Boulder.

4. Data Availability

Figure 1. September Monthly Average Arctic Sea Ice Extent, 1979–2012

Users can access monthly map images, GIS-compatible map files, and gridded daily and monthly satellite data, along with corresponding metadata, at: http://nsidc.org/data/seaice_index/archives/index.html. From this page, users can also download monthly extent and area data. From this page, select "Get Extent and Concentration Data," which will lead to a public FTP site (ftp://sidads.colorado.edu/DATASETS/NOAA/G02135). To obtain the September monthly data that were used in this indicator, select the "Sep" directory, then choose the "...area.txt" file with the data. To see a different version of the graph in Figure 1 (plotting percent anomalies rather than square miles), return to the "Sep" directory and open the "...plot.png" image.

NSIDC's Sea Ice Index documentation page (http://nsidc.org/data/docs/noaa/g02135_seaice_index) describes how to download, read, and interpret the data. It also defines database fields and key terminology. Gridded source data can be found at: http://nsidc.org/data/nsidc-0051.html and: http://nsidc.org/data/nsidc-0081.html.

Figure 2. Age of Arctic Sea Ice at Minimum September Week, 1983–2012

NSIDC published a version of Figure 2 at: http://nsidc.org/arcticseaicenews/2012/10/poles-apart-a-record-breaking-summer-and-winter. EPA obtained the data shown in the figure by contacting NSIDC. The data are processed by Dr. James Maslanik and Dr. Mark Tschudi at the University of Colorado, Boulder, and provided to NSIDC. Earlier versions of this analysis appeared in Maslanik et al. (2011) and Maslanik et al. (2007).

Satellite data used in historical and ongoing monitoring of sea ice age can be found at the following websites:

- Defense Meteorological Satellite Program (DMSP) Scanning Multi Channel Microwave Radiometer (SMMR): http://nsidc.org/data/nsidc-0071.html
- DMSP Special Sensor Microwave/Imager (SSM/I): http://nsidc.org/data/nsidc-0001.html
- DMSP Special Sensor Microwave Imager and Sounder (SSMIS): http://nsidc.org/data/nsidc-0001.html
- NASA Advanced Microwave Scanning Radiometer for the Earth Observing System (AMSR-E): http://nsidc.org/data/amsre
- Advanced Very High Resolution Radiometer (AVHRR): http://nsidc.org/data/avhrr/data_summaries.html

Age calculations also depend on wind measurements and on buoy-based measurements and, motion vectors. Wind measurements are available at: www.esrl.noaa.gov/psd/data/reanalysis/reanalysis.shtml. Data and metadata are available online at: http://iabp.apl.washington.edu/data.html and: http://nsidc.org/data/nsidc-0116.html.

Methodology

5. Data Collection

This indicator is based on maps of sea ice extent in the Arctic Ocean and surrounding waters, which were developed using brightness temperature imagery collected by satellites. Data from October 1978 through June 1987 were collected using the Nimbus-7 SMMR instrument, and data since July 1987 have been collected using a series of successor SSM/I instruments. In 2008, the SSMIS replaced the SSM/I as the source for sea ice products. These instruments can identify the presence of sea ice because sea ice and open water have different passive microwave signatures. The record has been supplemented with data from AMSR-E, which operated from 2003 to 2011.

The satellites that supply data for this indicator orbit the Earth continuously, collecting images that can be used to generate daily maps of sea ice extent. They are able to map the Earth's surface with a resolution of 25 kilometers. The resultant maps have a nominal pixel area of 625 square kilometers.

Because of the curved map projection, however, actual pixel sizes range from 382 to 664 square kilometers.

The satellites that collect the data cover most of the Arctic region in their orbital paths. However, the sensors cannot collect data from a circular area immediately surrounding the North Pole due to orbit inclination. From 1978 through June 1987, this "pole hole" measured 1.19 million square kilometers. Since July 1987 it has measured 0.31 million square kilometers. For more information about this spatial gap and how it is corrected in the final data, see Section 6.

To calculate the age of ice (Figure 2), the SSM/I, SMMR, and AMSR-E imagery have been supplemented with three additional data sets:

- AVHRR satellite data, which use an optical sensing instrument that can measure sea ice temperature and heat flux, which in turn can be used to estimate thickness. AVHRR also covers the "pole hole."
- Maps of wind speed and direction at 10 meters above the Earth's surface, which were compiled by NOAA's National Centers for Environmental Prediction (NCEP).
- Motion vectors that trace how parcels of sea ice move, based on data collected by the International Arctic Buoy Programme (IABP). Since 1955, the IABP has deployed a network of 14 to 30 *in situ* buoys in the Arctic Ocean that provide information about movement rates at six-hour intervals.

For documentation of passive microwave satellite data collection methods, see the summary and citations at: http://nsidc.org/data/docs/noaa/g02135_seaice_index. For further information on AVHRR imagery, see: http://noaasis.noaa.gov/NOAASIS/ml/avhrr.html. For motion tracking methods, see Maslanik et al. (2011), Fowler et al. (2004), and: http://nsidc.org/data/nsidc-0116.html.

6. Indicator Derivation

Figure 1. September Monthly Average Arctic Sea Ice Extent, 1979–2012

Satellite data are used to develop daily ice extent and concentration maps using an algorithm developed by NASA. Data are evaluated within grid cells on the map. Image processing includes quality control features such as two weather filters based on brightness temperature ratios to screen out false positives over open water, an ocean mask to eliminate any remaining sea ice in regions where sea ice is not expected, and a coastal filter to eliminate most false positives associated with mixed land/ocean grid cells.

From each daily map, analysts calculate the total "extent" and "area" covered by ice. These terms are defined differently as a result of how they address those portions of the ocean that are partially but not completely frozen:

- **Extent** is the total area covered by all pixels on the map that have at least 15 percent ice concentration, which means at least 15 percent of the ocean surface within that pixel is frozen over. The 15 percent concentration cutoff for extent is based on validation studies that showed that a 15 percent threshold provided the best approximation of the "true" ice edge and the lowest bias. In practice, most of the area covered by sea ice in the Arctic far exceeds the 15 percent threshold, so using a higher cutoff (e.g., 20 or 30 percent) would yield different totals but similar overall trends (for example, see Parkinson et al., 1999).

- **Area** represents the actual surface area covered by ice. If a pixel's area were 600 square kilometers and its ice concentration were 75 percent, then the ice area for that pixel would be 450 square kilometers. At any point in time, total ice area will always be less than total ice extent.

EPA's indicator addresses extent rather than area. Both of these measurements are valid ways to look at trends in sea ice, but in this case, EPA chose to look at the time series for extent because it is more complete than the time series for area. In addition, the available area data set does not include the "pole hole" (the area directly above the North Pole that the satellites cannot cover), and the size of this unmapped region changed as a result of the instrumentation change in 1987, creating a discontinuity in the area data. In contrast, the extent time series assumes that the entire "pole hole" area is covered with at least 15 percent ice, which is a reasonable assumption based on other observations of this area.

NASA's processing algorithm includes steps to deal with occasional days with data gaps due to satellite or sensor outages. These days were removed from the time series and replaced with interpolated values based on the total extent of ice on the surrounding days.

From daily maps and extent totals, NSIDC calculated monthly average extent in square kilometers. EPA converted these values to square miles to make the results accessible to a wider audience. By relying on monthly averages, this indicator smoothes out some of the variability inherent in daily measurements.

Figure 1 shows trends in September average sea ice extent. September is when Arctic sea ice typically reaches its annual minimum, after melting during the summer months. By looking at the month with the smallest extent of sea ice, this indicator focuses attention on the time of year when limiting conditions would most affect wildlife and human societies in the Arctic region.

This indicator does not attempt to estimate values from before the onset of regular satellite mapping in October 1978 (which makes 1979 the first year with September data for this indicator). It also does not attempt to project data into the future.

For documentation of the NASA Team algorithm used to process the data, see Cavalieri et al. (1984) and: http://nsidc.org/data/nsidc-0051.html. For more details about NSIDC methods, see the Sea Ice Index documentation and related citations at: http://nsidc.org/data/docs/noaa/g02135_seaice_index.

Other months of the year were considered for this indicator, but EPA chose to focus on September, which is when the extent of ice reaches its annual minimum. September extent is often used as an indicator. One reason is because as temperatures start to get colder, there may be less meltwater on the surface than during the previous summer months, thus leading to more reliable remote sensing of ice extent. Increased melting during summer months leads to changes in the overall character of the ice—i.e., age and thickness—and these changes have implications throughout the year. Thus, September conditions are particularly important for the overall health of Arctic sea ice.

Evidence shows that the extent of Arctic sea ice has declined in all months of the year. Comiso (2012) examined the seasonal pattern in Arctic sea ice extent for three decadal periods plus the years 2007, 2009, and 2010 and found declines throughout the year. The figure below shows an analysis of monthly means from the National Snow and Ice Data Center (NSIDC)—the source of data for this indicator. It reveals that Arctic sea ice extent has declined in all months, with the most pronounced decline in the summer and fall.

Monthly Arctic Sea Ice Extent,
1978/1979 - 2011/2012

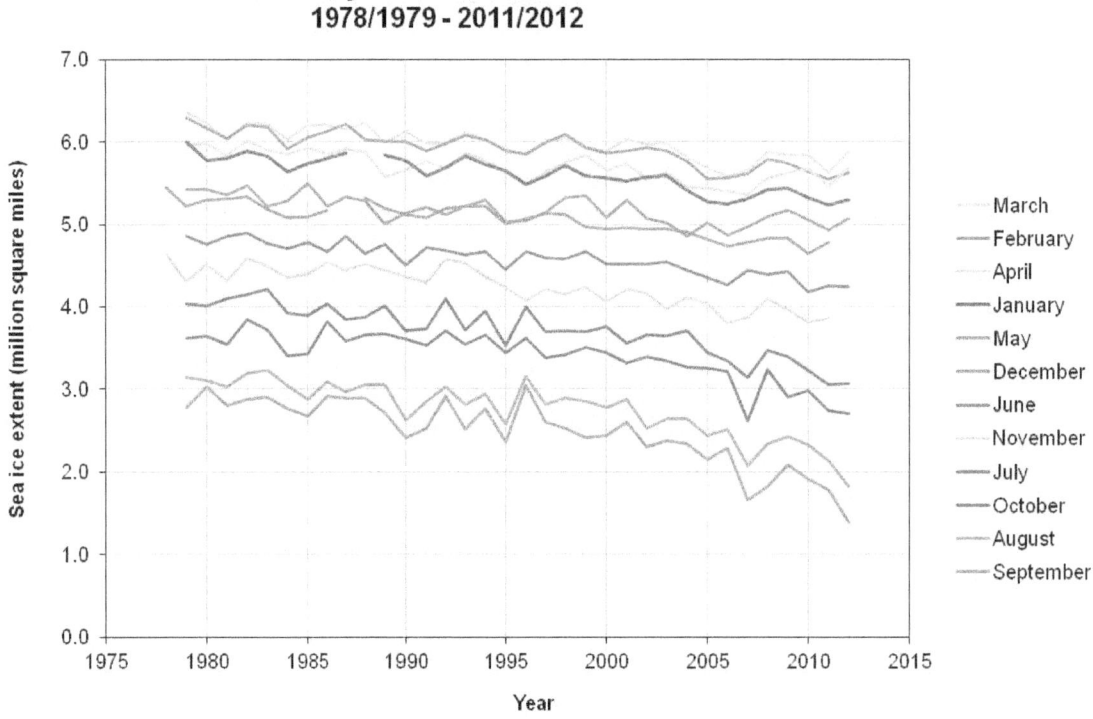

Data source: NSIDC: http://nsidc.org/data/seaice_index/archives/index.html. *Accessed November 2012.*

Figure 2. Age of Arctic Sea Ice at Minimum September Week, 1983–2012

A research team at the University of Colorado at Boulder processes daily sequential SSM/I, SMMR, AMSR-E, and AVHRR satellite data from NASA, then produces maps using a grid with 12 km-by-12 km cells. The AVHRR data help to fill the "pole hole" and provide information about the temperature and thickness of the ice. Like Figure 1, this method classifies a pixel as "ice" if at least 15 percent of the ocean surface within the area is frozen over. Using buoy data from the IABP, motion vectors for the entire region are blended via optimal interpolation and mapped on the gridded field. NCEP wind data are also incorporated at this stage, with lower weighting during winter and higher weighting during summer, when surface melt limits the performance of the passive microwave data. Daily ice extent and motion vectors are averaged on a weekly basis. Once sea ice reaches its annual minimum extent (typically in early September), the ice is documented as having aged by one year. For further information on data processing methods, see Maslanik et al. (2011), Maslanik et al. (2007), and Fowler et al. (2004). Although the most recently published representative study does not utilize AMSR-E brightness data or NCEP wind data for the calculation of ice motion, the results presented in Figure 2 and the NSIDC website incorporate these additional sources.

Figure 2 shows the extent of ice that falls into several age categories. Whereas Figure 1 extends back to 1979, Figure 2 can only show trends back to 1983 because it is not possible to know how much ice is five or more years old (the oldest age class shown) until parcels of ice have been tracked for at least five years. Regular satellite data collection did not begin until October 1978, which makes 1983 the first year in which September minimum ice can be assigned to the full set of age classes shown in Figure 2.

7. Quality Assurance and Quality Control

Image processing includes a variety of quality assurance and quality control (QA/QC) procedures, including steps to screen out false positives. These procedures are described in NSIDC's online documentation at: http://nsidc.org/data/docs/noaa/g02135_seaice_index as well as in some of the references cited therein.

NSIDC Arctic sea ice data have three levels of processing for quality control. NSIDC's most recent data come from the Near Real-Time SSM/I Polar Gridded Sea Ice Concentrations (NRTSI) data set. NRTSI data go through a first level of calibration and quality control to produce a "PRELIM" preliminary data product. The final data are processed by NASA's Goddard Space Flight Center (GSFC), which uses a similar process but a higher level of QC. Switching from NRTSI to GSFC data can result in slight changes in the total extent values—on the order of 50,000 square kilometers or less for total sea ice extent.

Because PRELIM and GSFC processing requires several months' lag time, Figure 1 reports GSFC data for the years 1979 to 2010 and a NRTSI data point for 2011. At the time EPA published this report, the GSFC data for 2011 had not yet been finalized.

Analysis

8. Comparability Over Time and Space

Both figures for this indicator are based on data collection methods and processing algorithms that have been applied consistently over time and space. NASA's satellites cover the entire area of interest with the exception of a "hole" at the North Pole for Figure 1. Even though the size of this hole has changed over time, EPA's indicator uses a data set that corrects for this discontinuity.

The total extent shown in Figure 2 (the sum of all the stacked areas) differs from the total extent in Figure 1 because Figure 2 shows conditions during the specific week in September when minimum extent is reached, while Figure 1 shows average conditions over the entire month of September. It would not make sense to convert Figure 2 to a monthly average for September because all ice is "aged" one year as soon as the minimum has been achieved, which creates a discontinuity after the minimum week.

9. Sources of Uncertainty

NSIDC's Sea Ice Index documentation (http://nsidc.org/data/docs/noaa/g02135_seaice_index) describes several analyses that have examined the accuracy and uncertainty of passive microwave imagery and the NASA Team algorithm used to create this indicator. For example, a 1991 analysis estimated that ice concentrations measured by passive microwave imagery are accurate to within 5 to 9 percent, depending on the ice being imaged. Another study suggested that the NASA Team algorithm underestimates ice extent by 4 percent in the winter and more in summer months. A third study that compared the NASA Team algorithm with new higher-resolution data found that the NASA Team algorithm underestimates ice extent by an average of 10 percent. For more details and study citations, see: http://nsidc.org/data/docs/noaa/g02135_seaice_index. Certain types of ice conditions can lead to larger errors, particularly thin or melting ice. For example, a melt pond on an ice floe might be mapped as open water. The instruments also can have difficulty distinguishing the interface between ice and snow or a diffuse boundary between ice and open water. Using the September minimum minimizes many of these effects because melt ponds and the ice surface become largely frozen by then. These errors do not affect trends and relative changes from year to year.

NSIDC has calculated standard deviations along with each monthly ice concentration average.

NSIDC has considered using a newer algorithm that would process the data with greater certainty, but doing so would require extensive research and reprocessing, and data from the original instrument (pre-1987) might not be compatible with some of the newer algorithms that have been proposed. Thus, for the time being, this indicator uses the best available science to provide a multi-decadal representation of trends in Arctic sea ice extent. The overall trends shown in this indicator have been corroborated by numerous other sources, and readers should feel confident that the indicator provides an accurate overall depiction of trends in Arctic sea ice over time.

Accuracy of ice motion vectors depends on the error in buoy measurements, wind fields, and satellite images. Given that buoy locational readings are taken every six hours, satellite images are 24-hour averages, and a "cm/sec" value is interpolated based on these readings, accuracy depends on the error of the initial position and subsequent readings. NSIDC proposes that "the error would be less than 1 cm/sec for the average velocity over 24 hours" (http://nsidc.org/data/docs/daac/nsidc0116_icemotion/buoy.html).

10. Sources of Variability

Many factors contribute to variability in this indicator. In constructing the indicator, several choices have been made to minimize the extent to which this variability affects the results. The apparent extent of sea ice can vary widely from day to day, both due to real variability in ice extent (growth, melting, and movement of ice at the edge of the ice pack) and due to ephemeral effects such as weather, clouds and water vapor, melt on the ice surface, and changes in the character of the snow and ice surface. The intensity of Northern Annular Mode (NAM) conditions and changes to the Arctic Oscillation also have a strong year-to-year impact on ice movement. Under certain conditions, older ice might move to warmer areas subject to increased melting. Weather patterns can also affect the sweeping of icebergs out of the Arctic entirely. For a more complete description of major thermodynamic processes that impact ice longevity, see Maslanik et al. (2007) and Rigor and Wallace (2004).

According to NSIDC's documentation at: http://nsidc.org/data/docs/noaa/g02135_seaice_index, extent is a more reliable variable than ice concentration or area. The weather and surface effects described above can substantially impact estimates of ice concentration, particularly near the edge of the ice pack. Extent is a more stable variable because it simply registers the presence of at least a certain percentage of sea ice in a grid cell (15 percent). For example, if a particular pixel has an ice concentration of 50 percent, outside factors could cause the satellite to measure the concentration very differently, but as long as the result is still greater than the percent threshold, this pixel will be correctly accounted for in the total "extent." Monthly averages also help to reduce some of the day-to-day "noise" inherent in sea ice measurements.

11. Statistical/Trend Analysis

This indicator does not report on the slope of the apparent trends in September sea ice extent and age distribution, nor does it calculate the statistical significance of these trends.

12. Data Limitations

Factors that may impact the confidence, application, or conclusions drawn from this indicator are as follows:

1. Variations in sea ice are not entirely due to changes in temperature. Other conditions, such as fluctuations in oceanic and atmospheric circulation and typical annual and decadal variability, can also affect the extent of sea ice, and by extension the sea ice age indicator.
2. Changes in the age and thickness of sea ice—for example, a trend toward younger or thinner ice—might increase the rate at which ice melts in the summer, making year-to-year comparisons more complex.
3. Many factors can diminish the accuracy of satellite mapping of sea ice. Although satellite instruments and processing algorithms have improved somewhat over time, applying these new methods to established data sets can lead to trade-offs in terms of reprocessing needs and compatibility of older data. Hence, this indicator does not use the highest-resolution imagery or the newest algorithms. Trends are still accurate, but should be taken as a general representation of trends in sea ice extent, not an exact accounting.
4. As described in Section 6, the threshold used to determine extent—15 percent ice cover within a given pixel—represents an arbitrary cutoff without a particular scientific significance. Nonetheless, studies have found that choosing a different threshold would result in a similar overall trend. Thus, the most important part of Figure 1 is not the absolute extent reported for any given year, but the size and shape of the trend over time.
5. Using ice surface data and motion vectors only allows the determination of a maximum sea ice age. Thus, as presented, the Figure 2 indicator only indicates the age distribution of sea ice on the surface, and is not necessarily representative of the age distribution of the total sea ice volume.

References

Cavalieri, D.J., P. Gloersen, and W.J. Campbell. 1984. Determination of sea ice parameters with the NIMBUS-7 SMMR. J. Geophys. Res. 89(D4):5355–5369.

Comiso, J. 2012. Large decadal decline of the Arctic multiyear ice cover. J. Climate 25(4):1176–1193.

Fowler, C., W.J. Emery, and J. Maslanik. 2004. Satellite-derived evolution of Arctic sea ice age: October 1978 to March 2003. IEEE Geosci. Remote Sens. Lett. 1(2):71–74.

Maslanik, J.A., C. Fowler, J. Stroeve, S. Drobot, J. Zwally, D. Yi, and W. Emery. 2007. A younger, thinner Arctic ice cover: Increased potential for rapid, extensive sea-ice loss. Geophys. Res. Lett. 34:L24501.

Maslanik, J., J. Stroeve, C. Fowler, and W. Emery. 2011. Distribution and trends in Arctic sea ice age through spring 2011. Geophys. Res. Lett. 38:L13502.

Parkinson, C.L., D.J. Cavalieri, P. Gloersen, H.J. Zwally, and J.C. Comiso. 1999. Arctic sea ice extents, areas, and trends, 1978–1996. J. Geophys. Res. 104(C9):20,837–20,856.

Rigor, I.G., and J.M. Wallace. 2004. Variations in the age of Arctic sea-ice and summer sea-ice extent. Geophys. Res. Lett. 31:L09401. http://iabp.apl.washington.edu/research_seaiceageextent.html.

Glaciers

Identification

1. Indicator Description

This indicator examines the balance between snow accumulation and melting in glaciers, and describes how the size of glaciers around the world has changed since 1945. On a local and regional scale, changes in glaciers have implications for ecosystems and people who depend on glacier-fed streamflow. On a global scale, loss of ice from glaciers contributes to sea level rise.

Components of this indicator include:

- Cumulative trends in the mass balance of reference glaciers worldwide over the past 65 years (Figure 1)
- Cumulative trends in the mass balance of three U.S. glaciers over the past half-century (Figure 2)

2. Revision History

April 2010: Indicator posted
December 2011: Updated with data through 2010
April 2012: Replaced Figure 1 with data from a new source
June 2012: Updated Figure 2 with data through 2010 for South Cascade Glacier

Data Sources

3. Data Sources

Figure 1 shows the average cumulative mass balance of a global set of reference glaciers, which was originally published by the World Glacier Monitoring Service (WGMS) (2011). Measurements were collected by a variety of academic and government programs and compiled by WGMS.

The U.S. Geological Survey (USGS) Benchmark Glacier Program provided the data for Figure 2, which shows the cumulative mass balance of three U.S. "benchmark" glaciers where long-term monitoring has taken place.

4. Data Availability

Figure 1. Average Cumulative Mass Balance of "Reference Glaciers" Worldwide, 1945–2010

A version of Figure 1 with data through 2009 was published in WGMS (2011). Preliminary values for 2010 were posted by WGMS at: www.wgms.ch/mbb/sum10.html. As a result, 2010 is associated with a reduced number (30) of associated reference glaciers. EPA obtained the data in spreadsheet form from the staff of WGMS.

Raw measurements of glacier surface parameters around the world have been recorded in a variety of formats. Some data are available in online databases such as the World Glacier Inventory (http://nsidc.org/data/glacier_inventory/index.html). Some raw data are also available in studies by USGS. WGMS (www.geo.uzh.ch/microsite/wgms/) maintains perhaps the most comprehensive record of international observations—much of it in hard copy only.

Figure 2. Cumulative Mass Balance of Three U.S. Glaciers, 1958–2010

A cumulative net mass balance data set is available on the USGS benchmark glacier website at: http://ak.water.usgs.gov/glaciology/all_bmg/3glacier_balance.htm. Because the online data are not necessarily updated every time a correction or recalculation is made, EPA obtained the most up-to-date data for Figure 2 directly from USGS. More detailed metadata and measurements from the three benchmark glaciers can be found on the USGS website at: http://ak.water.usgs.gov/glaciology.

Methodology

5. Data Collection

This indicator provides information on the cumulative change in mass balance of numerous glaciers over time. Glacier mass balance data are calculated based on a variety of measurements at the surface of a glacier, including measurements of snow depths and snow density. These measurements help glaciologists determine changes in snow and ice accumulation and ablation that result from snow precipitation, snow compaction, freezing of water, melting of snow and ice, calving (i.e., ice breaking off from the tongue or leading edge of the glacier), wind erosion of snow, and sublimation from ice (Mayo et al., 2004). Both surface size and density of glaciers are measured to produce net mass balance data. These data are reported in meters of water equivalent (mwe), which corresponds to the average change in thickness over the entire surface area of the glacier. Because snow and ice can vary in density (depending on the degree of compaction, for example), converting to the equivalent amount of liquid water provides a more consistent metric.

Measurement techniques have been described and analyzed in many peer-reviewed studies, including Josberger et al. (2007). Most long-term glacier observation programs began as part of the International Geophysical Year in 1957–1958.

Figure 1. Average Cumulative Mass Balance of "Reference Glaciers" Worldwide, 1945–2010

The global trend is based on data collected at 37 reference glaciers around the world, which are as follows:

Continent	Region	Glaciers
North America	Alaska	Gulkana, Wolverine
North America	Pacific Coast Ranges	Place, South Cascade, Helm, Lemon Creek, Peyto
North America	Canadian High Arctic	Devon Ice Cap NW, Meighen Ice Cap, White
South America	Andes	Echaurren Norte

Continent	Region	Glaciers
Europe	Svalbard	Austre Broeggerbreen, Midtre Lovénbreen
Europe	Scandinavia	Engabreen, Alfotbreen, Nigardsbreen, Grasubreen, Storbreen, Hellstugubreen, Hardangerjoekulen, Storglaciaeren
Europe	Alps	Saint Sorlin, Sarennes, Argentière, Silvretta, Gries, Stubacher Sonnblickkees, Vernagtferner, Kesselwandferner, Hintereisferner, Caresèr
Europe/Asia	Caucasus	Djankuat
Asia	Altai	No. 125 (Vodopadniy), Maliy Aktru, Leviy Aktru
Asia	Tien Shan	Ts. Tuyuksuyskiy, Urumqi Glacier No.1

WGMS chose these 37 reference glaciers because they all had at least 30 years of continuous mass balance records (WGMS, 2011). As the small graph at the bottom of Figure 1 shows, some of these glaciers have data extending as far back as the 1940s. WGMS did not include data from glaciers that are dominated by non-climatic factors, such as surge dynamics or calving. Because of data availability and the distribution of glaciers worldwide, WGMS's compilation is dominated by the Northern Hemisphere.

All of the mass balance data that WGMS compiled for this indicator are based on the direct glaciological method (Østrem and Brugman, 1991), which involves manual measurements with stakes and pits at specific points on each glacier's surface.

Figure 2. Cumulative Mass Balance of Three U.S. Glaciers, 1958–2010

Figure 2 shows data collected at the three glaciers studied by USGS's Benchmark Glacier Program. All three glaciers have been monitored for many decades. USGS chose them because they represent typical glaciers found in their respective regions: South Cascade Glacier in the Pacific Northwest (a continental glacier), Wolverine Glacier in coastal Alaska (a maritime glacier), and Gulkana Glacier in inland Alaska (a continental glacier). Hodge et al. (1998) and Josberger et al. (2007) provide more information about the locations of these glaciers and why USGS selected them for the benchmark monitoring program.

USGS collected repeated measurements at each of the glaciers to determine the various parameters that can be used to calculate cumulative mass balance. Specific information on sampling design at each of the three glaciers is available in Bidlake et al. (2010) and Van Beusekom et al. (2010). Measurements are collected at specific points on the glacier surface, designated by stakes.

Data for South Cascade Glacier are available beginning in 1959 (relative to conditions in 1958) and for Gulkana and Wolverine Glaciers beginning in 1966 (relative to conditions in 1965). Glacier monitoring methodology has evolved over time based on scientific re-analysis, and cumulative net mass balance data for these three glaciers are routinely updated as glacier measurement methodologies improve and more information becomes available. Several papers that document data updates through time are available on the USGS benchmark glacier website at: http://ak.water.usgs.gov/glaciology.

6. Indicator Derivation

For this indicator, glacier surface measurements have been used to determine the net change in mass balance from one year to the next, referenced to the previous year's summer surface measurements. The indicator documents changes in mass and volume rather than total mass or volume of each glacier because the latter is more difficult to determine accurately. Thus, the indicator is not able to show how the magnitude of mass balance change relates to the overall mass of the glacier (e.g., what percentage of the glacier's mass has been lost).

Glaciologists convert surface measurements to mass balance by interpolating measurements over the glacier surface geometry. Two different interpolation methods can be used: conventional balance and reference-surface balance. In the conventional balance method, measurements are made at the glacier each year to determine glacier surface geometry, and other measurements are interpolated over the annually modified geometry. The reference-surface balance method does not require that glacier geometry be re-determined each year. Rather, glacier surface geometry is determined once, generally the first year that monitoring begins, and the same geometry is used each of the following years. A more complete description of conventional balance and reference-surface balance methods is given in Harrison et al. (2009).

Mass balance is typically calculated over a balance year, which begins at the onset of snow and ice accumulation. For example, the balance year at Gulkana Glacier starts and ends in September of each year. Thus, the balance year beginning in September 2010 and ending in September 2011 is called "balance year 2011." Annual mass balance changes are confirmed based on measurements taken the following spring.

Figure 1. Average Cumulative Mass Balance of "Reference Glaciers" Worldwide, 1945–2010

The graph shows the average cumulative mass balance of WGMS's reference glaciers over time. The number of reference glaciers included in this calculation varies by year, but it is still possible to generate a reliable time series because the figure shows an average across all of the glaciers measured, rather than a sum. No attempt was made to extrapolate from the observed data in order to calculate a cumulative global change in mass balance.

Figure 2. Cumulative Mass Balance of Three U.S. Glaciers, 1958–2010

At each of the three benchmark glaciers, changes in mass balance have been summed over time to determine the cumulative change in mass balance since a reference year. For the sake of comparison, all three glaciers use a reference year of 1965, which is set to zero. Thus, a negative value in a later year means the glacier has lost mass since 1965. All three time series in Figure 2 reflect the conventional mass balance method, as opposed to the reference-surface method. No attempt has been made to project the results for the three benchmark glaciers to other locations. See Bidlake et al. (2010), Van Beusekom et al. (2010), and sources cited therein for further description of analytical methods.

As noted in the report (in the caption and in the figure), the data point for South Cascade Glacier is preliminary. USGS provided the 2010 data point for South Cascade Glacier with the following caveat:

> Material presented here is preliminary in nature and presented prior to final review. The data and information are provided with the understanding that they are not guaranteed to be correct or complete. Users are cautioned to consider carefully the provisional nature of these data and information before using them for decisions that concern personal or public safety or the conduct of business that involves substantial monetary or operational consequences. Conclusions drawn from, or actions undertaken on the basis of, such data and information are the sole responsibility of the user.

USGS formally "approves" annual mass balance estimates at five-year intervals, per their publication schedule. However, they can still provide new data to support annual updates of this indicator, as long as the new data points are labeled appropriately. EPA chose to include the 2010 South Cascade Glacier data point because it has been reviewed by USGS and revised in response to follow-up measurements. EPA elected not to include 2011 mass balance numbers because they are still subject to adjustment in response to follow-up measurements.

7. Quality Assurance and Quality Control

The underlying measurements for Figure 1 come from a variety of data collection programs, each with its own procedures for quality assurance and quality control (QA/QC). WGMS also has its own requirements for data quality. For example, WGMS only incorporates measurements that reflect the direct glaciological method (Østrem and Brugman, 1991).

USGS periodically reviews and updates the mass balance data shown in Figure 2. For example, in Fountain et al. (1997), the authors explain that mass balance should be periodically compared with changes in ice volume, as the calculations of mass balance are based on interpolation of point measurements that are subject to error. In addition, March (2003) describes steps that USGS takes to check the weighting of certain mass balance values. This weighting allows USGS to convert point values into glacier-averaged mass balance values.

Ongoing reanalysis of glacier monitoring methods, described in several of the reports listed on USGS's website (http://ak.water.usgs.gov/glaciology), provides an additional level of quality control for data collection.

Analysis

8. Comparability Over Time and Space

Glacier monitoring methodology has evolved over time based on scientific re-analysis of methodology. Peer-reviewed studies describing the evolution of glacier monitoring are listed in Mayo et al. (2004). Figure 2 accounts for these changes, as USGS periodically reanalyzes past data points using improved methods.

The reference glaciers tracked in Figure 1 reflect a variety of methods over time and space, and it is impractical to adjust for all of these small differences. However, as a general indication of trends in glacier mass balance, Figure 1 shows a clear pattern whose strength is not diminished by the inevitable variety of underlying sources.

9. Sources of Uncertainty

Glacier measurements have inherent uncertainties. For example, maintaining a continuous and consistent data record is difficult because the stakes that denote measurement locations are often distorted by glacier movement and snow and wind loading. Additionally, travel to measurement sites is dangerous and inclement weather can prevent data collection during the appropriate time frame. In a cumulative time series, such as the analyses presented in this indicator, the size of the margin of error grows with time because each year's value depends on all of the preceding years.

Figure 1. Average Cumulative Mass Balance of "Reference Glaciers" Worldwide, 1945–2010

Uncertainties have been quantified for some glacier mass balance measurements, but not for the combined time series shown in Figure 1. WGMS (2011) has identified greater quantification of uncertainty in mass balance measurements as a key goal for future research.

Figure 2. Cumulative Mass Balance of Three U.S. Glaciers, 1958–2010

Annual mass balance measurements for the three USGS benchmark glaciers usually have an estimated error of ±0.1 to ±0.2 meters of water equivalent (Josberger et al., 2007). Error bars for the two Alaskan glaciers are plotted in Van Beusekom et al. (2010). Further information on error estimates is given in Bidlake et al. (2010) and Van Beusekom et al. (2010). Harrison et al. (2009) describe error estimates related to interpolation methods.

10. Sources of Variability

Glacier mass balance can reflect year-to-year variations in temperature, precipitation, and other factors. Figure 2 shows some of this year-to-year variability, while Figure 1 shows less variability because the change in mass balance has been averaged over many glaciers around the world. In both cases, the availability of several decades of data allows the indicator to show long-term trends that exceed the "noise" produced by interannual variability. In addition, the period of record is longer than the period of key multi-year climate oscillations such as the Pacific Decadal Oscillation and El Niño–Southern Oscillation, meaning the trends shown in Figures 1 and 2 are not simply the product of decadal-scale climate oscillations.

11. Statistical/Trend Analysis

Figures 1 and 2 both show a cumulative loss of mass or volume over time, from which analysts can derive an average annual rate of change. Confidence bounds are not provided for the trends in either figure, although both Bidlake et al. (2010) and Van Beusekom et al. (2010) cite clear evidence of a decline in mass balance at U.S. benchmark glaciers over time.

12. Data Limitations

Factors that may impact the confidence, application, or conclusions drawn from this indicator are as follows:

1. Slightly different methods of measurement and interpolation have been used at different glaciers, making direct year-to-year comparisons of change in cumulative net mass balance or volume difficult. Overall trends among glaciers can be compared, however.
2. The number of glaciers with data available to calculate mass balance in Figure 1 decreases as one goes back in time. Thus, averages from the 1940s to the mid-1970s rely on a smaller set of reference glaciers than the full 37 compiled in later years.
3. The relationship between climate change and glacier mass balance is complex, and the observed changes at a specific glacier might reflect a combination of global and local climate variations.
4. Records are available from numerous other individual glaciers in the United States, but many of these other records lack the detail, consistency, or length of record provided by the USGS

benchmark glaciers program. USGS has collected data on these three glaciers for decades using consistent methods, and USGS experts suggest that at least a 30-year record is necessary to provide meaningful statistics. Due to the complicated nature of glacier behavior, it is difficult to assess the significance of observed trends over shorter periods (Josberger et al., 2007).

References

Bidlake, W.R., E.G. Josberger, and M.E. Savoca. 2010. Modeled and measured glacier change and related glaciological, hydrological, and meteorological conditions at South Cascade Glacier, Washington, balance and water years 2006 and 2007. U.S. Geological Survey Scientific Investigations Report 2010–5143. http://pubs.usgs.gov/sir/2010/5143/.

Fountain, A.G., R.M. Krimmel, and D.C. Trabant. 1997. A strategy for monitoring glaciers. U.S. Geological Survey Circular 1132.

Harrison, W.D., L.H. Cox, R. Hock, R.S. March, and E.C. Petit. 2009. Implications for the dynamic health of a glacier from comparison of conventional and reference-surface balances. Ann. Glaciol. 50:25–30.

Hodge, S.M., D.C. Trabant, R.M. Krimmel, T.A. Heinrichs, R.S. March, and E.G. Josberger. 1998. Climate variations and changes in mass of three glaciers in western North America. J. Climate 11:2161–2217.

Josberger, E.G., W.R. Bidlake, R.S. March, and B.W. Kennedy. 2007. Glacier mass-balance fluctuations in the Pacific Northwest and Alaska, USA. Ann. Glaciol. 46:291–296.

March, R.S. 2003. Mass balance, meteorology, area altitude distribution, glacier-surface altitude, ice motion, terminus position, and runoff at Gulkana Glacier, Alaska, 1996 balance year. U.S. Geological Survey Water-Resources Investigations Report 03-4095.

Mayo, L.R., D.C. Trabant, and R.S. March. 2004. A 30-year record of surface mass balance (1966–95) and motion and surface altitude (1975–95) at Wolverine Glacier, Alaska. U.S. Geological Survey Open-File Report 2004-1069.

Østrem, G., and M. Brugman. 1991. Glacier mass-balance measurements: A manual for field and office work. National Hydrology Research Institute (NHRI), NHRI Science Report No. 4.

Van Beusekom, A.E., S.R. O'Neel, R.S. March, L.C. Sass, and L.H. Cox. 2010. Re-analysis of Alaskan benchmark glacier mass-balance data using the index method. U.S. Geological Survey Scientific Investigations Report 2010–5247. http://pubs.usgs.gov/sir/2010/5247/

WGMS (World Glacier Monitoring Service). 2011. Glacier mass balance bulletin no. 11 (2008–2009). Zemp, M., S.U. Nussbaumer, I. Gärtner-Roer, M. Hoelzle, F. Paul, and W. Haeberli (eds.). ICSU(WDS)/IUGG(IACS)/UNEP/UNESCO/WMO. Zurich, Switzerland: World Glacier Monitoring Service. www.wgms.ch/mbb/mbb11/wgms_2011_gmbb11.pdf.

Lake Ice

Identification

1. Indicator Description

This indicator measures the amount of time that ice is present on lakes in the United States between approximately 1850 and 2010. If lakes remain frozen for longer periods, it can signify that the climate is cooling. Conversely, shorter periods of ice cover suggest a warming climate.

Components of this indicator include:

- Trends in the duration of ice cover on selected U.S. lakes since 1850 (Figure 1)
- Trends in first freeze dates of selected U.S. lakes since 1850 (Figure 2)
- Trends in ice breakup dates of selected U.S. lakes since 1850 (Figure 3)

2. Revision History

April 2010: Indicator posted
December 2011: Updated with data through winter 2010–2011

Data Sources

3. Data Sources

This indicator is based on data from the Global Lake and River Ice Phenology Database, which was compiled by the North Temperate Lakes Long-Term Ecological Research program at the Center for Limnology at the University of Wisconsin–Madison from data submitted by participants in the Lake Ice Analysis Group (LIAG). The database is hosted on the Web by the National Snow and Ice Data Center (NSIDC), and it currently contains ice cover data for 750 lakes and rivers throughout the world, some with records as long as 150 years.

4. Data Availability

All of the lake ice observations used for this indicator are publicly available from NSIDC's Global Lake and River Ice Phenology Database. Users can access this database at: http://nsidc.org/data/lake_river_ice. Database documentation can be found at: http://nsidc.org/data/docs/noaa/g01377_lake_river_ice.

Users can also view descriptive information about each lake or river in the Global Lake and River Ice Phenology Database. The Global Lake and River Ice Phenology Database contains the following fields, although many records are incomplete:

- Lake or river name
- Lake or river code
- Whether it is a lake or a river

- Continent
- Country
- State
- Latitude (decimal degrees)
- Longitude (decimal degrees)
- Elevation (meters)
- Mean depth (meters)
- Maximum depth (meters)
- Median depth (meters)
- Surface area (square kilometers)
- Shoreline length (kilometers)
- Largest city population
- Power plant discharge (yes or no)
- Area drained (square kilometers)
- Land use code (urban, agriculture, forest, grassland, other)
- Conductivity (microsiemens per centimeter)
- Secchi depth (Secchi disk depth in meters)
- Contributor

Access to the Global Lake and River Ice Phenology Database is unrestricted, but users are encouraged to register so they can receive notification of changes to the database in the future.

Methodology

5. Data Collection

This indicator examines three parameters related to ice cover on lakes:

- The annual "ice on" or freeze date, defined as the first date on which the water body was observed to be completely covered by ice.
- The annual "ice off," thaw, or breakup date, defined as the date of the last breakup observed before the summer open water phase.
- The annual duration of ice cover, defined as the number of days that a water body is completely covered with ice. If a lake thawed for several days in mid-winter and then froze again, the duration would equal the number of days from ice on to ice off minus those days when the lake thawed.

Observers have gathered data on lake ice throughout the United States for many years—in some cases, more than 100 years. The types of observers can vary from one location to another. For example, some observations might have been gathered and published by a local newspaper editor; others compiled by a local resident. Some lakes have benefited from multiple observers, such as residents on both sides of the lake who can compare notes to determine when the lake is completely frozen or thawed. At some locations, observers have kept records of all three parameters of interest ("ice on," "ice off," and total ice duration); others might have tracked only one or two of these parameters.

To ensure sound spatial and temporal coverage, EPA limited this indicator to U.S. water bodies with the longest and most complete historical records. After downloading data for all lakes and rivers within the United States, EPA sorted the data and analyzed each water body to determine data availability for the three parameters of interest. As a result of this analysis, EPA identified eight water bodies—all lakes— with particularly long and rich records. Special emphasis was placed on identifying water bodies with many consecutive years of data, which can support moving averages and other trend analysis. EPA selected the following eight lakes for trend analysis:

- Detroit Lake, Minnesota
- Lake George, New York
- Lake Mendota, Wisconsin
- Lake Michigan (Grand Traverse Bay), Michigan
- Lake Monona, Wisconsin
- Lake Otsego, New York
- Mirror Lake, New York
- Shell Lake, Wisconsin

Together, these lakes span much of the Great Lakes region and upstate New York.

6. Indicator Derivation

To smooth out some of the variability in the annual data and to make it easier to see long-term trends in the display, EPA did not plot annual time series but instead calculated nine-year moving averages (arithmetic means) for each of the parameters. EPA chose a nine-year period because it is consistent with other indicators and comparable to the 10-year moving averages used in a similar analysis by Magnuson et al. (2000). Average values are plotted at the center of each nine-year window. For example, the average from 1990 to 1998 is plotted at year 1994. EPA did calculate averages over periods that were missing a few data points. Early years sometimes had sparse data, and the earliest averages were calculated only around the time when many consecutive records started to appear in the record for a given lake.

EPA used endpoint padding to extend the nine-year smoothed lines all the way to the ends of the analysis period for each lake. For example, if annual data were available through 2010, EPA calculated smoothed values centered at 2007, 2008, 2009, and 2010 by inserting the 2006–2010 average into the equation in place of the as-yet-unreported annual data points for 2011 and beyond. EPA used an equivalent approach at the beginning of each time series.

For consistency, all data points in Figures 1, 2, and 3 are plotted at the base year, which is the year the winter season began. For the winter of 2010 to 2011, the base year would be 2010, even if a particular lake did not freeze until early 2011.

EPA did not attempt to interpolate missing data points and did not attempt to calculate duration in cases where only the ice on and ice off date were provided. Such manipulations would have been based on unfounded assumptions. This indicator also does not attempt to portray data beyond the time periods of observation or beyond the eight lakes that were selected for the analysis.

Magnuson et al. (2000) and Jensen et al. (2007) describe methods of processing lake ice observations for use in calculating long-term trends.

7. Quality Assurance and Quality Control

The LIAG performed some basic quality control checks on data that were contributed to the database, making corrections in some cases. Additional corrections continue to be made as a result of user comments. For a description of some recent corrections, see the database documentation at: http://nsidc.org/data/docs/noaa/g01377_lake_river_ice.

Ice observations rely on human judgment. Definitions of "ice on" and "ice off" vary, and the definitions used by any given observer are not necessarily documented alongside the corresponding data. Where possible, the scientists who developed the database have attempted to use sources that appear to be consistent from year to year, such as a local resident with a long observation record.

Analysis

8. Comparability Over Time and Space

Historical observations have not been made systematically or according to a standard protocol. Rather, the Global Lake and River Ice Phenology Database—the source of data for this indicator—represents a systematic effort to compile data from a variety of original sources.

All three parameters were determined by human observations that incorporate some degree of personal judgment. Definitions of the three parameters can also vary over time and from one location to another. Human observations provide an advantage, however, in that they enable trend analysis over a much longer time period than can be afforded by more modern techniques such as satellite imagery. Overall, human observations provide the best available record of seasonal ice formation and breakup, and the breadth of available data allows analysis of broad spatial patterns as well as long-term temporal patterns.

9. Sources of Uncertainty

Ice observations rely on human judgment, and definitions of "ice on" and "ice off" vary, which could lead to some uncertainty in the data. For example, some observers might consider a lake to have thawed once they can no longer walk on it, while others might wait until the ice has entirely melted. Observations also depend on one's vantage point along the lake, particularly a larger lake—for example, if some parts of the lake have thawed while others remain frozen. In addition, the definitions used by any given observer are not necessarily documented alongside the corresponding data. Therefore, it is not possible to ensure that all variables have been measured consistently from one lake to another—or even at a single lake over time—and it is also not possible to quantify the true level of uncertainty or correct for such inconsistencies.

Accordingly, the Global Lake and River Ice Phenology Database does not provide error estimates for historical ice observations. Where possible, however, the scientists who developed the database have attempted to use sources that appear to be consistent from year to year, such as a local resident who collects data over a long period. Overall, the Global Lake and River Ice Phenology Database represents the best available data set for lake ice observations, and limiting the indicator to eight lakes with the most lengthy and complete records should lead to results in which users can have confidence.

10. Sources of Variability

For a general idea of the variability inherent in these types of time series, see Magnuson et al. (2000) and Jensen et al. (2007)—two papers that discuss variability and statistical significance for a broader set of lakes and rivers, including some of the lakes in this indicator. Magnuson et al. (2005) discuss variability between lakes, considering the extent to which observed variability reflects factors such as climate patterns, lake morphometry (shape), and lake trophic status. The timing of freeze-up and break-up of ice appears to be more sensitive to air temperature changes at lower latitudes (Livingstone et al., 2010), but despite this, lakes at higher latitudes appear to be experiencing the most rapid reductions in duration of ice cover (Latifovic and Pouliot, 2007).

To smooth out some of the interannual variability and to make it easier to see long-term trends in the display, EPA did not plot annual time series but instead calculated nine-year moving averages (arithmetic means) for each of the parameters, following an approach recommended by Magnuson et al. (2000).

11. Statistical/Trend Analysis

Figures 1, 2, and 3 show trends for each of the eight lakes. No attempt was made to aggregate the eight lakes together. EPA calculated trends over time by ordinary least-squares regression, a common statistical method, to support some of the statements in the "Key Points" section of the indicator. EPA has not calculated the statistical significance of these particular long-term trends, although Magnuson et al. (2000) and Jensen et al. (2007) found that long-term trends in freeze and breakup dates for many lakes were statistically significant (p<0.05).

12. Data Limitations

Factors that may impact the confidence, application, or conclusions drawn from this indicator are as follows:

1. Although the Global Lake and River Ice Phenology Database provides a lengthy historical record of freeze and thaw dates for a much larger set of lakes and rivers, some records are incomplete, ranging from brief lapses to large gaps in data. Thus, this indicator is limited to eight lakes with relatively complete historical records. Geographic coverage is limited to sites in four states (Minnesota, Wisconsin, Michigan, and New York).

2. Data used in this indicator are all based on visual observations. Records based on visual observations by individuals are open to some interpretation and can reflect different definitions and methods.

3. Historical observations for lakes have typically been made from the shore, which might not be representative of lakes as a whole or comparable to more recent satellite-based observations.

References

Jensen, O.P., B.J. Benson, and J.J. Magnuson. 2007. Spatial analysis of ice phenology trends across the Laurentian Great Lakes region during a recent warming period. Limnol. Oceanogr. 52(5):2013–2026.

Latifovic, R., and D. Pouliot. 2007. Analysis of climate change impacts on lake ice phenology in Canada using the historical satellite data record. Remote Sens. Environ. 106:492–507.

Livingstone, D.M., R. Adrian, T. Blencker, G. George, and G.A. Weyhenmeyer. 2010. Lake ice phenology. In: George, D.G. (ed). The impact of climate change on European lakes. Aquatic Ecology Series 4:51–62.

Magnuson, J., D. Robertson, B. Benson, R. Wynne, D. Livingstone, T. Arai, R. Assel, R. Barry, V. Card, E. Kuusisto, N. Granin, T. Prowse, K. Steward, and V. Vuglinski. 2000. Historical trends in lake and river ice cover in the Northern Hemisphere. Science 289:1743–1746.

Magnuson, J.J., B.J. Benson, O.P. Jensen, T.B. Clark, V. Card, M.N. Futter, P.A. Soranno, and K.M. Stewart. 2005. Persistence of coherence of ice-off dates for inland lakes across the Laurentian Great Lakes region. Verh. Internat. Verein. Limnol. 29:521–527.

Snowfall

Identification

1. Indicator Description

Warmer temperatures associated with climate change can influence snowfall by altering weather patterns, causing more precipitation overall, and causing more precipitation to fall in the form of rain instead of snow. This indicator examines how snowfall has changed across the contiguous 48 states over time.

Components of this indicator include:

- Trends in total winter snowfall accumulation in the contiguous 48 states since 1930 (Figure 1)
- Changes in the ratio of snowfall to total winter precipitation since 1949 (Figure 2)

2. Revision History

December 2011: Indicator developed
May 2012: Updated Figure 2 with data through 2011

Data Sources

3. Data Sources

The data used for this indicator are based on two studies published in the peer-reviewed literature: Kunkel et al. (2009) (Figure 1) and a 2012 update to Feng and Hu (2007) (Figure 2). Both studies are based on long-term weather station records compiled by the National Oceanic and Atmospheric Administration's (NOAA's) National Climatic Data Center (NCDC).

4. Data Availability

Figure 1. Change in Total Snowfall in the Contiguous 48 States, 1930–2007

EPA acquired Figure 1 data directly from Dr. Kenneth Kunkel of NOAA's Cooperative Institute for Climate and Satellites (CICS). Kunkel's analysis is based on data from weather stations that are part of NOAA's Cooperative Observer Program (COOP). Complete data, embedded definitions, and data descriptions for these stations can be found online at: www.ncdc.noaa.gov/doclib/. State-specific data can be found at: www7.ncdc.noaa.gov/IPS/coop/coop.html;jsessionid=312EC0892FFC2FBB78F63D0E3ACF6CBC. There are no confidentiality issues that may limit accessibility. Additional metadata can be found at: www.nws.noaa.gov/om/coop/.

Figure 2. Change in Snow-to-Precipitation Ratio in the Contiguous 48 States, 1949–2005

EPA acquired data for this indicator from Dr. Song Feng of the University of Nebraska, Lincoln, based on results published in Feng and Hu (2007) and updated in 2012. Underlying data come from the U.S. Historical Climatology Network (USHCN), a compilation of weather station data maintained by NOAA. The USHCN allows users to download daily or monthly data at: www.ncdc.noaa.gov/oa/climate/research/ushcn/. This website also provides data descriptions and other metadata. The data were taken from USHCN Version 2.

Methodology

5. Data Collection

Systematic collection of weather data in the United States began in the 1800s. Since then, observations have been recorded from 23,000 different stations. At any given time, observations are recorded from approximately 8,000 stations.

NOAA's National Weather Service (NWS) operates some stations (called first-order stations), but the vast majority of U.S. weather stations are part of the COOP network, which represents the core climate network of the United States (Kunkel et al., 2005). Cooperative observers include state universities, state and federal agencies, and private individuals. Observers are trained to collect data following NWS protocols, and equipment to gather these data is provided and maintained by the NWS.

Data collected by COOP are referred to as U.S. Daily Surface Data or Summary of the Day data. General information about the NWS COOP data set is available at: www.nws.noaa.gov/os/coop/what-is-coop.html. Sampling procedures are described in the full metadata for the COOP data set available at: www.nws.noaa.gov/om/coop/.

NCDC also maintains the USHCN, which contains data from a subset of COOP and first-order weather stations that meet certain selection criteria and undergo additional levels of quality control. USHCN contains precipitation data from approximately 1,200 stations within the contiguous 48 states. The period of record varies for each station but generally includes most of the 20th century. One of the objectives in establishing the USHCN was to detect secular changes of regional rather than local climate. Therefore, stations included in this network are only those believed to not be influenced to any substantial degree by artificial changes of local environments. To be included in the USHCN, a station had to meet certain criteria for record longevity, data availability (percentage of available values), spatial coverage, and consistency of location (i.e., experiencing few station changes). An additional criterion, which sometimes compromised the preceding criteria, was the desire to have a uniform distribution of stations across the United States. Included with the data set are metadata files that contain information about station moves, instrumentation, observing times, and elevation. NOAA's website (www.ncdc.noaa.gov/oa/climate/research/ushcn) provides more information about USHCN data collection.

Figure 1. Change in Total Snowfall in the Contiguous 48 States, 1930–2007

The analysis in Figure 1 is based on snowfall (in inches), which weather stations measure daily through manual observation using a snow measuring rod. The study on which this indicator is based includes data from 419 COOP stations in the contiguous United States for the months of October to May. These stations were selected using screening criteria that were designed to identify stations with the most consistent methods and most reliable data over time. Screening criteria are described in greater detail in Section 7.

Figure 2. Change in Snow-to-Precipitation Ratio in the Contiguous 48 States, 1949–2011

The analysis in Figure 2 is based primarily on snowfall and precipitation measurements collected with standard gauges that "catch" precipitation, thus allowing weather stations to report daily precipitation totals. This study uses data from 289 USHCN stations in the contiguous United States. Stations south of 37°N latitude were not included because most of them receive minimal amounts of snow each year. Additional site selection criteria are described in Section 7. This analysis covers the months from November through March, and each winter has been labeled based on the year in which it ends. For example, the data for "1949" represent the season that extended from November 1948 through March 1949.

6. Indicator Derivation

Figure 1. Change in Total Snowfall in the Contiguous 48 States, 1930–2007

At each station, daily snowfall totals have been summed to get the total snowfall for each winter. Thus, this figure technically reports trends from the winter of 1930–1931 to the winter of 2006–2007. Long-term trends in snowfall accumulation for each station are derived using an ordinary least-squares linear regression of the annual totals. Kunkel et al. (2009) describe analytical procedures in more detail.

Figure 2. Change in Snow-to-Precipitation Ratio in the Contiguous 48 States, 1949–2011

Using precipitation records from the USHCN Version 2, the liquid-water equivalent of daily snowfall was calculated according to methods described by Huntington et al. (2004) and Knowles et al. (2006). For each station, a snow-to-precipitation (S:P) ratio was calculated for each year by comparing the total snowfall during the months of interest (in terms of liquid-water equivalent) with total precipitation (snow plus rain). Long-term rates of change at each station were determined using a Kendall's tau slope estimator. This method of statistical analysis is described in Sen (1968) and Gilbert (1987). For a more detailed description of analytical methods, see Feng and Hu (2007).

7. Quality Assurance and Quality Control

The NWS has documented COOP methods, including training manuals and maintenance of equipment, at: www.nws.noaa.gov/os/coop/training.htm. These training materials also discuss quality control of the underlying data set. Additionally, pre-1948 data in the COOP data set have recently been digitized from hard copy. Quality control procedures associated with digitization and other potential sources of error are discussed in Kunkel et al. (2005).

Quality control procedures for USHCN Version 1 are summarized at: www.ncdc.noaa.gov/oa/climate/research/ushcn/ushcn.html#QUAL. Homogeneity testing and data correction methods are described in numerous peer-reviewed scientific papers by NCDC. Quality control procedures for USHCN Version 2 are summarized at: www.ncdc.noaa.gov/oa/climate/research/ushcn/#processing.

Figure 1. Change in Total Snowfall in the Contiguous 48 States, 1930–2007

Kunkel et al. (2009) filtered stations for data quality by selecting stations with records that were at least 90 percent complete over the study period. In addition, each station must possess at least five years of records during the decade at either end of the trend analysis (i.e., 1930s and 2000s) because data near the endpoints exert a relatively heavy influence on the overall trend. Year-to-year statistical outliers were also extensively cross-checked against nearby stations or *Climatological Data* publications when available. Any discrepancies with apparent regional trends were reviewed and evaluated by a panel of seven climate experts for data quality assurance. A more extensive description of this process, along with other screening criteria, can be found in Kunkel et al. (2009).

Figure 2. Change in Snow-to-Precipitation Ratio in the Contiguous 48 States, 1949–2011

Feng and Hu (2007) applied a similar filtering process to ensure data quality and consistency over time. Stations missing certain amounts of snow or precipitation data per month or per season were excluded from the study. Additional details about quality assurance are described in Feng and Hu (2007).

The 2012 update to Feng and Hu (2007) added another screening criterion that excluded sites that frequently used a particular estimation method to calculate snow water equivalent. This resulted in 85 fewer stations in the 2012 data set. Specifically, instructions given to observers in the early to mid-twentieth century provided an option to convert the measured snowfall to precipitation using a 10:1 ratio if it was impractical to melt the snow. Many observers have used this option in their reports of daily precipitation, although the number of observers using this option has declined through the years. The actual snowfall to liquid precipitation ratio is related to the air temperature during the snow event, and it also varies spatially. The median ratio in recent decades has been approximately 13:1 in the contiguous United States (Baxter et al., 2005; Kunkel et al., 2007), which suggests that using a 10:1 ratio could generally overestimate daily precipitation. Total winter precipitation in a snowy climate would thus be problematic if a large portion of the daily precipitation was estimated using this ratio. To reduce the impact of this practice on cold season precipitation, the 2012 analysis excluded records where winter (November to March) had more than 10 days with snowfall depth larger than 3.0 cm and where more than 50 percent of those snowy days reported total precipitation using the 10:1 ratio.

Analysis

8. Comparability Over Time and Space

Techniques for measuring snow accumulation and precipitation are comparable over space and time, as are the analytical methods that were used to develop Figures 1 and 2. Steps have been taken to remove stations where trends could be biased by changes in methods, location, or surrounding land cover.

9. Sources of Uncertainty

Quantitative estimates of uncertainty are not available for Figure 1, Figure 2, or most of the underlying measurements.

Figure 1. Change in Total Snowfall in the Contiguous 48 States, 1930–2007

Snow accumulation measurements are subject to human error. Despite the vetting of observation stations, some error could also result from the effects of wind and surrounding cover, such as tall trees. Some records have evidence of reporting error related to missing data (i.e., days with no snow being reported as missing data instead of "0 inches"), but Kunkel et al. (2009) took steps to correct this error in cases where other evidence (such as daily temperatures) made it clear that an error was present.

Figure 2. Change in Snow-to-Precipitation Ratio in the Contiguous 48 States, 1949–2011

The source study classifies all precipitation as "snow" for any day that received some amount of snow. This approach has the effect of overestimating the amount of snow during mixed snow-sleet-rain conditions. Conversely, wind effects that might prevent snow from settling in gauges will tend to bias the S:P ratio toward rainier conditions.

10. Sources of Variability

Snowfall naturally varies from year to year as a result of typical variation in weather patterns, multi-year climate cycles such as the El Niño–Southern Oscillation and Pacific Decadal Oscillation, and other factors. However, both components of this indicator cover more than 50 years of data, thus allowing for a reliable analysis of long-term trends.

Snowfall is influenced by temperature and a host of other factors such as regional weather patterns, local elevation and topography, and proximity to large water bodies. These differences can lead to great variability in trends among stations—even stations that may be geographically close to one another.

11. Statistical/Trend Analysis

Figure 1. Change in Total Snowfall in the Contiguous 48 States, 1930–2007

This indicator reports a trend for each station based on ordinary least-squares linear regression. The significance of each station's trend has not been reported.

Figure 2. Change in Snow-to-Precipitation Ratio in the Contiguous 48 States, 1949–2011

Feng and Hu (2007) calculated a long-term trend in S:P ratio at each station using the Kendall's tau method. The same method was used for the 2012 update. The authors also determined a z-score for every station. Based on these z-scores, Figure 2 identifies which station trends are statistically significant based on a 95 percent confidence threshold (i.e., a z-score with an absolute value greater than 1.645).

12. Data Limitations

Factors that may impact the confidence, application, or conclusions drawn from this indicator are as follows:

1. Several factors make it difficult to measure snowfall precisely. The snow accumulations shown in Figure 1 are based on the use of measuring rods. This measurement method is subject to human error, as well as the effects of wind (drifting snow) and the surrounding environment (such as tall trees). Similarly, snow gauges for Figure 2 may catch slightly less snow than rain because of

the effects of wind. However, steps have been taken to limit this indicator to weather stations with the most consistent methods and the highest-quality data.

2. Both figures are limited to the winter season. Figure 1 comes from an analysis of October-to-May snowfall, while Figure 2 covers November through March. Although these months account for the vast majority of snowfall in most locations, this indicator might not represent the entire snow season in some areas.

3. Taken by itself, a decrease in S:P ratio does not necessarily mean that a location is receiving less snow than it used to or that snow has changed to rain. For example, a station with increased rainfall in November might show a decline in S:P ratio even with no change in snowfall during the rest of the winter season. This example illustrates the value of examining snowfall trends from multiple perspectives, as this indicator seeks to do.

4. Selecting only those stations with high-quality long-term data leads to an uneven density of stations for this indicator. Low station density limits the conclusions that can be drawn about certain regions such as the Northeast and the Intermountain West.

References

Baxter, M.A., C.E. Graves, and J.T. Moore. 2005. A climatology of snow-to-liquid ratio for the contiguous United States. Weather and Forecasting 20:729–744.

Feng, S., and Q. Hu. 2007. Changes in winter snowfall/precipitation ratio in the contiguous United States. J. Geophys. Res. 112:D15109.

Gilbert, R.O. 1987. Statistical methods for environmental pollution monitoring. New York, NY: Van Nostrand Reinhold.

Huntington, T.G., G.A. Hodgkins, B.D. Keim, and R.W. Dudley. 2004. Changes in the proportion of precipitation occurring as snow in New England (1949–2000). J. Clim. 17:2626–2636.

Knowles, N., M.D. Dettinger, and D.R. Cayan. 2006. Trends in snowfall versus rainfall in the western United States. J. Clim. 19:4545–4559.

Kunkel, K.E., D.R. Easterling, K. Hubbard, K. Redmond, K. Andsager, M.C. Kruk, and M.L. Spinar. 2005. Quality control of pre-1948 Cooperative Observer Network data. J. Atmos. Oceanic Technol. 22:1691–1705.

Kunkel, K.E., M. Palecki, K.G. Hubbard, D.A. Robinson, K.T. Redmond, and D.R. Easterling. 2007. Trend identification in twentieth-century U.S. snowfall: The challenges. J. Atmos. Oceanic Technol. 24:64–73.

Kunkel, K.E., M. Palecki, L. Ensor, K.G. Hubbard, D. Robinson, K. Redmond, and D. Easterling. 2009. Trends in twentieth-century U.S. snowfall using a quality-controlled dataset. J. Atmos. Oceanic Technol. 26:33–44.

Sen, P.K. 1968. Estimates of the regression coefficient based on Kendall's tau. J. Am. Stat. Assoc. 63:1379–1389.

Snow Cover

Identification

1. Indicator Description

This indicator measures changes in the amount of land in North America covered by snow. The amount of land covered by snow at any given time is influenced by climate factors such as the amount of snowfall an area receives, the timing of that snowfall, and the rate of melting on the ground.

Components of this indicator include:

- Average annual snow cover since 1972 (Figure 1)
- Average snow cover by season since 1972 (Figure 2)

2. Revision History

April 2010: Indicator posted
January 2012: Updated with data through 2011
February 2012: Expanded to include snow cover by season (new Figure 2)

Data Sources

3. Data Sources

This indicator is based on a Rutgers University Global Snow Lab (GSL) reanalysis of digitized maps produced by the National Oceanic and Atmospheric Administration (NOAA) using their Interactive Multisensor Snow and Ice Mapping System (IMS).

4. Data Availability

Complete weekly and monthly snow cover extent data for North America (excluding Greenland) are publicly available for users to download from the GSL website at: http://climate.rutgers.edu/snowcover/table_area.php?ui_set=2. A complete description of these data can be found on the GSL website at: http://climate.rutgers.edu/snowcover/index.php.

The underlying NOAA gridded maps are also publicly available. To obtain these maps, visit the NOAA IMS website at: www.natice.noaa.gov/ims.

Methodology

5. Data Collection

This indicator is based on data from satellite instruments. These satellites orbit the Earth continuously, collecting images that can be used to generate weekly maps of snow cover. Data are collected for the entire Northern Hemisphere; this indicator includes data for all of North America, excluding Greenland.

Data were compiled as part of NOAA's IMS, which incorporates imagery from a variety of satellite instruments (Advanced Very High Resolution Radiometer [AVHRR], Geostationary Satellite Server [GOES], Special Sensor Microwave Imager [SSMI], etc.) as well as derived mapped products and surface observations. Characteristic textured surface features and brightness allow for snow to be identified and data to be collected on percent of snow cover and surface albedo (Robinson et al., 1993).

NOAA's IMS website (www.natice.noaa.gov/ims) lists peer-reviewed studies that discuss the data collection methods. For example, NOAA sampling procedures are described in Ramsay (1998).

6. Indicator Derivation

NOAA digitizes satellite maps weekly using the National Meteorological Center Limited-Area Fine Mesh grid. In the digitization process, an 89-by-89-cell grid is placed over the Northern Hemisphere and each cell has a resolution range of 16,000 to 42,000 square kilometers. NOAA then analyzes snow cover within each of these grid cells.

Rutgers University's GSL reanalyzes the digitized maps produced by NOAA to correct for biases in the data set caused by locations of land masses and bodies of water that NOAA's land mask does not completely resolve. Initial reanalysis produces a new set of gridded data points based on the original NOAA data points. Both original NOAA data and reanalyzed data are filtered using a more detailed land mask produced by GSL. These filtered data are then used to make weekly estimates of snow cover. GSL determines the weekly extent of snow cover by placing an 89-by-89-cell grid over the Northern Hemisphere snow cover map and calculating the total area of all grid cells that are at least 50 percent snow-covered. To generate monthly maps, GSL weights weekly areas based on the number of days of each week that fall within a given month.

EPA obtained weekly estimates of snow-covered area and averaged them to determine the annual average extent of snow cover in square kilometers. EPA obtained monthly estimates of snow-covered area to determine the seasonal extent of snow cover in square kilometers. For each year, a season's extent was determined by averaging the following months:

- Winter: December (of the prior calendar year), January, and February
- Spring: March, April, and May
- Summer: June, July, and August
- Fall: September, October, and November

EPA converted all of these values to square miles to make the results accessible to a wider audience.

NOAA's IMS website describes the initial creation and digitization of gridded maps; see: www.natice.noaa.gov/ims. The GSL website provides a complete description of how GSL reanalyzed NOAA's gridded maps to determine weekly and monthly snow cover extent. See: http://climate.rutgers.edu/snowcover/docs.php?target=vis and http://climate.rutgers.edu/snowcover/docs.php?target=cdr. Robinson et al. (1993) describe GSL's methods, while Helfrich et al. (2007) document how GSL has accounted for methodological improvements over time.

7. Quality Assurance and Quality Control

Quality assurance and quality control (QA/QC) measures occur throughout the analytical process, most notably in the reanalysis of NOAA data by GSL. GSL's filtering and correction steps are described online (http://climate.rutgers.edu/snowcover/docs.php?target=vis) and in Robinson et al. (1993). Ramsey (1998) describes the validation plan for NOAA digitized maps and explains how GSL helps to provide objective third party verification of NOAA data.

Analysis

8. Comparability Over Time and Space

Steps have been taken to exclude less reliable early data from this indicator. Although NOAA satellites began collecting snow cover imagery in 1966, early maps had a lower resolution than later maps (4 kilometers versus 1 kilometer in later maps) and the early years also had many weeks with missing data. Data collection became more consistent with better resolution in 1972, when a new instrument called the Very High Resolution Radiometer (VHRR) came online. This indicator only presents data from 1972 and later.

Mapping methods have continued to evolve since 1972. Accordingly, GSL has taken steps to reanalyze older maps to ensure consistency with the latest approach. GSL provides more information about these correction steps at: http://climate.rutgers.edu/snowcover/docs.php?target=cdr.

Data have been collected and analyzed using consistent methods over space. The satellites that collect the data cover all of North America in their orbital paths.

9. Sources of Uncertainty

Uncertainty measurements are not readily available for this indicator or for the underlying data. Although exact uncertainty estimates are not available, extensive QA/QC and third-party verification measures show that steps have been taken to minimize uncertainty and ensure that users are able to draw accurate conclusions from the data. Documentation available from GSL (http://climate.rutgers.edu/snowcover/docs.php?target=vis) explains that since 1972, satellite mapping technology has had sufficient accuracy to support continental-scale climate studies. Although satellite data have some limitations (see Section 12), maps based on satellite imagery are often still superior to maps based on ground observations, which can be biased due to the preferred position of weather stations in valleys and in places affected by urban heat islands, such as airports. Hence, satellite-based maps are generally more representative of regional snow extent, particularly for mountainous or sparsely populated regions.

10. Sources of Variability

Figures 1 and 2 show substantial year-to-year variability in snow cover. This variability naturally results from variation in weather patterns, multi-year climate cycles such as the El Niño–Southern Oscillation and Pacific Decadal Oscillation, and other factors. Underlying weekly measurements have even more variability. This indicator accounts for these factors by presenting a long-term record (several decades) and calculating annual and seasonal averages.

Generally, decreases in snow cover duration have been most pronounced along mid-latitude continental margins where seasonal mean air temperatures range from -5 to +5°C (Brown and Mote, 2009).

11. Statistical/Trend Analysis

Upon the advice of experts from GSL, EPA did not attempt to define trends using a single linear regression. Instead, EPA determined ranges and decadal averages to support some of the statements in the "Key Points." Decadal averages suggest that the extent of snow cover has declined over time.

12. Data Limitations

Factors that may impact the confidence, application, or conclusions drawn from this indicator are as follows:

1. Satellite data collection is limited by anything that obscures the ground, such as low light conditions at night, dense cloud cover, or thick forest canopy. Satellite data are also limited by difficulties discerning snow cover from other similar-looking features such as cloud cover.
2. Although satellite-based snow cover totals are available starting in 1966, some of the early years are missing data from several weeks (mainly during the summer), which would lead to an inaccurate annual or seasonal average. Thus, the indicator is restricted to 1972 and later, with all years having a full set of data.
3. Summer snow mapping is particularly complicated because many of the patches of snow that remain (e.g., high in a mountain range) are smaller than the pixel size for the analysis. This leads to reduced confidence in summer estimates. When summer values are incorporated into an annual average, however, variation in summer values has relatively minimal influence on the overall results.

References

Brown, R.D., and P.W. Mote. 2009. The response of Northern Hemisphere snow cover to a changing climate. J. Climate 22:2124–2145.

Helfrich, S.R., D. McNamara, B.H. Ramsay, T. Baldwin, and T. Kasheta. 2007. Enhancements to, and forthcoming developments in the Interactive Multisensor Snow and Ice Mapping System (IMS). Hydrol. Process. 21:1576–1586.

Ramsay, B.H. 1998. The Interactive Multisensor Snow and Ice Mapping System. Hydrol. Process. 12:1537–1546.

Robinson, D.A., K.F. Dewey, and R.R. Heim, Jr. 1993. Global snow cover monitoring: An update. Bull. Am. Meteorol. Soc. 74:1689–1696.

Snowpack

Identification

1. Indicator Description

This indicator describes changes in springtime mountain snowpack in western North America between 1950 and 2000. Mountain snowpack is a key component of the water cycle in western North America, storing water in the winter when the snow falls and releasing it in spring and early summer when the snow melts.

2. Revision History

April 2010: Indicator posted

Data Sources

3. Data Sources

This indicator is based largely on data collected by the U.S. Department of Agriculture's (USDA's) Natural Resources Conservation Service (NRCS). Additional snowpack data come from observations made by the California Department of Water Resources and the British Columbia Ministry of the Environment. The analysis was developed by the authors of Mote et al. (2005).

4. Data Availability

EPA obtained the data for this indicator from Dr. Philip Mote at Oregon State University. Dr. Mote had published an earlier version of this analysis (Mote et al., 2005) with trends from 1950 through 1997, and he was able to provide EPA with an updated analysis of trends from 1950 through 2000.

Dr. Mote's analysis is based on snowpack measurements from NRCS, the British Columbia Ministry of the Environment, and the California Department of Water Resources. All three sets of data are available to the public with no confidentiality or accessibility restrictions. NRCS data are available at: www.wcc.nrcs.usda.gov/snow/snowhist.html. California data are available at: http://cdec.water.ca.gov/snow/current/snow/index2.html, and snowpack data for British Columbia are available at: http://a100.gov.bc.ca/pub/mss. These websites also provide descriptions of the data.

Methodology

5. Data Collection

This indicator uses snow water equivalent (SWE) measurements to assess trends in snowpack from 1950 to 2000. SWE is the amount of water contained within the snowpack at a particular location. It can be thought of as the depth of water that would result if the entire snowpack were to melt. Because snow can vary in density (depending on the degree of compaction, for example), converting to the equivalent amount of liquid water provides a more consistent metric.

Snowpack data have been collected over the years using a combination of manual and automated techniques. All of these techniques are ground-based observations, as snow water equivalent is difficult to measure from aircraft or satellites—although development and validation of remote sensing for snowpack is a subject of ongoing research. Consistent manual measurements from "snow courses" or observation sites are available beginning in the 1930s. In 1980, measurements began to be collected using an automated snowpack telemetry (SNOTEL) system, a set of remote sites that automatically measure snowpack and related climatic data. Snowpack measurements have been extensively documented and have been used for many years to help forecast spring and summer water supplies, particularly in the western United States.

The NRCS SNOTEL network operates over 650 remote sites in the western United States, including Alaska. Data from the SNOTEL network are augmented by manual snow course measurements. Manual snow course measurements are made monthly, while SNOTEL sensor data are recorded every 15 minutes and reported daily to two master stations.

Additional snowpack data come from observations made by the California Department of Water Resources and the British Columbia Ministry of the Environment.

For information about each of the data sources and its corresponding sample design, visit the following websites:

- NRCS: www.wcc.nrcs.usda.gov/snow/snowhist.html
- California Department of Water Resources: http://cdec.water.ca.gov/snow/info/DataCollecting.html
- British Columbia Ministry of the Environment: www.env.gov.bc.ca/rfc/data

The NRCS website describes both manual and telemetric snowpack measurement techniques in more detail at: www.wcc.nrcs.usda.gov/factpub/sect_4b.html. A training and reference guide for snow surveyors who use sampling equipment to measure snow accumulation is also available on the NRCS website at: www.wcc.nrcs.usda.gov/factpub/ah169/ah169.htm.

For consistency, this indicator examines trends at the same point in time each year. This indicator uses April 1st as the annual date for analysis because it is the most frequent observation date and it is extensively used for spring stream flow forecasting (Mote et al., 2005). Data are nominally attributed to April 1st, but in reality, for some manually operated sites the closest measurement in a given year might have been collected slightly before or after April 1st.

This indicator focuses on the western United States (excluding Alaska) and southwestern Canada because this broad region has the greatest density of stations with long-term records. A total of 1,155 locations have recorded SWE measurements within the area of interest. This indicator is based on 799 stations with sufficient April 1st records spanning the period from 1950 through 2000.

6. Indicator Derivation

Linear trends in April 1st SWE measurements were calculated for the period from 1950 through 2000 at each snow course or SNOTEL location, then these trends were converted to percent change since 1950. Note that this method can lead to an apparent loss exceeding 100 percent at a few sites (i.e., more than a 100 percent decrease in snowpack) in cases where the line of best fit passes through zero sometime before 2000, indicating that it is now most likely for that location to have no snowpack on the ground at all on April 1st. For more details about the analytical procedures used to calculate trends and percent change for each location, see Mote et al. (2005).

EPA obtained a data file with coordinates and percent change for each station, and plotted the results on a map using ArcGIS software. Figure 1 shows trends at individual sites with measured data, with no attempt to generalize data over space.

7. Quality Assurance and Quality Control

Automated SNOTEL data are screened by computer to ensure that they meet minimum requirements before being added to the database. In addition, each automated data collection site receives maintenance and sensor adjustment annually. Data reliability is verified by ground truth measurements taken during regularly scheduled manual surveys, in which manual readings are compared with automated data to check that values are consistent. Based on these quality assurance and quality control (QA/QC) procedures, maintenance visits are conducted to correct deficiencies. Additional description of QA/QC procedures for the SNOTEL network can be found on the NRCS website at: www.wcc.nrcs.usda.gov/factpub/sect_4b.html.

QA/QC procedures for manual measurements by NRCS do not appear to be available online. Details concerning QA/QC of data collected separately by California and British Columbia do not appear to be publicly available either.

Analysis

8. Comparability Over Time and Space

For consistency, this indicator examines trends at the same point in time each year. This indicator uses April 1st as the annual date for analysis because it is the most frequent observation date and it is extensively used for spring stream flow forecasting (Mote et al., 2005). Data are nominally attributed to April 1st, but in reality, for some manually operated sites the closest measurement in a given year might have been collected slightly before or after April 1st.

Data collection methods have changed over time in some locations, particularly as automated devices have replaced manual measurements. However, agencies such as NRCS have taken careful steps to calibrate the automated devices and ensure consistency between manual and automatic measurements (see Section 7). They also follow standard protocols to ensure that methods are applied consistently over time and space.

9. Sources of Uncertainty

Uncertainty estimates are not readily available for this indicator or for the underlying snowpack measurements. However, the regionally consistent and in many cases sizable changes shown in Figure 1 strongly suggest that this indicator shows real secular trends, not simply the artifacts of some type of measurement error.

10. Sources of Variability

Snowpack trends may be influenced by natural year-to-year variations in snowfall, temperature, and other climate variables. To reduce the influence of year-to-year variability, this indicator looks at longer-term trends over the full 51-year time series.

Over a longer timeframe, snowpack variability can result from multi-year oscillations in the Earth's climate or from nonclimatic factors such as changes in observation methods, land use, and forest canopy. Of particular note, the 1950s registered some of the highest snowpack measurements of the 20th century in the Northwest, and these high values at the start of the period of analysis could be magnifying the apparent snowpack decline depicted in Figure 1. However, further analysis reveals that the general direction of the trend is the same regardless of one's choice of start date.

11. Statistical/Trend Analysis

Figure 1 shows the results of a least-squares linear regression of annual observations at each individual site from 1950 to 2000. The statistical significance of each of these trends has not been reported.

12. Data Limitations

Factors that may impact the confidence, application, or conclusions drawn from this indicator are as follows:

1. EPA selected 1950 as a starting point for this analysis because data were readily available to examine trends throughout western North America from 1950 to present. Some others have looked at trends within smaller regions over longer or shorter timeframes, however, and found that the choice of start date can make a difference in the magnitude of the resulting trends. For example, Stoelinga et al. (2010) found a smaller long-term decline in snowpack in the Cascades when the analysis was extended back to 1930. This difference is due in part to several especially snowy years that occurred during the 1950s, which could be magnifying the extent of the snowpack decline depicted in Figure 1 for parts of the Northwest. However, evidence suggests that the general direction of the trend is the same regardless of the start date.
2. Although most parts of the West have seen reductions in snowpack, consistent with overall warming trends, observed snowfall trends could be partially influenced by non-climatic factors such as observation methods, land use changes, and forest canopy changes. A few snow course sites have been moved over time—for example, because of the growth of recreational uses such as snowmobiling or skiing. Mote et al. (2005) also report that the mean date of "April 1st" observations has grown slightly later over time.

References

Mote, P.W., A.F. Hamlet, M.P. Clark, and D.P. Lettenmaier. 2005. Declining mountain snowpack in western North America. Bull. Am. Meteorol. Soc. 86(1):39–49.

Stoelinga, M.T., M.D. Albright, and C.F. Mass. 2010. A new look at snowpack trends in the Cascade Mountains. J. Climate 23:2473–2491.

Streamflow

Identification

1. Indicator Description

This indicator describes trends in the magnitude and timing of streamflow in streams across the United States.

Components of this indicator include trends in three annual flow statistics:

- Magnitude of annual seven-day low streamflow from 1940 through 2009 (Figure 1)
- Magnitude of annual three-day high streamflow from 1940 through 2009 (Figure 2)
- Timing of winter-spring center of volume date from 1940 through 2009 (Figure 3)

2. Revision History

December 2011: Indicator developed
April 2012: Indicator updated with a new analysis

Data Sources

3. Data Sources

This indicator was developed by Drs. Mike McHale, Robert Dudley, and Glenn Hodgkins at the U.S. Geological Survey (USGS). It is based on streamflow data from a set of reference stream gauges specified in the Geospatial Attributes of Gages for Evaluating Streamflow (GAGES) database, which was developed by USGS and is described by Falcone et al. (2010). Daily mean streamflow data are stored in USGS's National Water Information System (NWIS).

4. Data Availability

EPA obtained the data for this indicator from Drs. Mike McHale, Robert Dudley, and Glenn Hodgkins at USGS. Similar streamflow analyses had been previously published in the peer-reviewed literature (Hodgkins and Dudley, 2006; Falcone et al., 2010). The USGS team provided a reprocessed dataset to include streamflow trends through 2009.

Streamflow data from individual stations are publicly available online through the surface water section of NWIS at: http://waterdata.usgs.gov/nwis/sw. Reference status and watershed, site characteristics, and other metadata for each stream gauge in the GAGES database are available online at: http://esapubs.org/archive/ecol/E091/045/.

Methodology

5. Data Collection

Streamflow is determined from data collected by devices called stream gauges, which record the elevation (or stage) of a river or stream at regular intervals each day. USGS maintains a national network of stream gauging stations, including more than 7,000 stations currently in operation throughout the contiguous 48 states (http://water.usgs.gov/wid/html/SG.html). USGS has been collecting stream gauge data since the late 1800s at some locations. However, gauges have not been not placed randomly; instead, they have been generally sited to capture information from relatively large perennial streams and rivers to record flows for specific management or legal issues. Stream surface elevation is recorded at regular intervals at each gauging station—typically every 15 minutes to 1 hour.

Streamflow (or discharge) is measured at regular intervals by USGS personnel (typically every four to eight weeks). The relation between stream surface elevation and discharge is determined and used to calculate streamflow for each stream stage measurement (Rantz et al., 1982). These data are used to calculate the daily mean discharge for each day at each site. All measurements are taken according to standard USGS procedures (Rantz et al., 1982; Sauer and Turnipseed, 2010; Turnipseed and Sauer, 2010).

This indicator uses data from a subset of USGS stream gauges that have been designated as "reference gauges" (Falcone et al., 2010). Reference gauges have been carefully selected to reflect minimal interference from human activities such as dam construction, reservoir management, wastewater treatment discharge, water withdrawals, and changes in land cover and land use that might influence runoff. The subset of reference gauges was further winnowed on the basis of length of period of record (70 years) and completeness of record (greater than or equal to 80 percent for every decade). Figures 1 and 2 are based on 211 sites. Figure 3 relies on 55 sites because it is limited to watersheds that receive 30 percent or more of their total annual precipitation in the form of snow. This additional criterion was applied because the metric in Figure 3 is used primarily to examine the timing of snowmelt-related runoff. All of the selected stations and their corresponding basins are independent—that is, the analysis does not include gauges that are upstream or downstream from one another.

All watershed characteristics, including basin area, station latitude and longitude, and percent of precipitation as snow were taken from the GAGES database. Basin area was determined through EPA's National Hydrography Dataset Plus and supplemented by the USGS National Water-Quality Assessment Program and the USGS Elevation Derivatives for National Applications.

6. Indicator Derivation

Figures 1 and 2. Volumes of Seven-Day Low (Figure 1) and Three-Day High (Figure 2) Streamflows in the United States, 1940–2009

Figure 1 shows trends in dry conditions using seven-day low streamflow, which is the lowest average of seven consecutive days of streamflow in a calendar year. Hydrologists commonly use this measure because it reflects sustained dry conditions that result in the lowest flows of the year. Seven-day low flow can equal zero if a stream has dried up completely.

Figure 2 shows trends in wet conditions using three-day high streamflow, which is the highest average of three consecutive days of streamflow in a calendar year. Hydrologists use this measure because a three-day averaging period has been shown to effectively characterize runoff associated with storms and peak snowmelt over a range of watershed areas.

Rates of change from 1940 to 2009 at each station were computed using the Sen slope, which is the median of all possible pair-wise slopes in a temporal dataset (Helsel and Hirsch, 1992). The Sen slope was then multiplied by the number of years of the trend period (i.e., 70) to estimate total change over time. Trends are reported as percentage increases or decreases, relative to the beginning Sen-slope value.

Figure 3. Timing of Winter-Spring Runoff in the United States, 1940–2009

Figure 3 shows trends in the timing of streamflow in the winter and spring, which is influenced by the timing of snowmelt runoff in areas with substantial annual snowpack. It does so using the winter-spring center of volume (WSCV) date, which is the date when half of the total volume of water between January 1 and May 31 has passed by the gauging station. Trends in this date are computed in the same manner as seven-day low flows and three-day high flows, and the results are reported in terms of the number of days earlier or later that WSCV is occurring. For more information about WSCV methods, see Hodgkins and Dudley (2006) and Burns et al. (2007).

7. Quality Assurance and Quality Control

Quality assurance and quality control (QA/QC) procedures are documented for measuring stream stage (Sauer and Turnipseed, 2010), measuring stream discharge (Turnipseed and Sauer, 2010), and computing stream discharge (Sauer, 2002; Rantz et al., 1982). Stream discharge is typically measured and equipment is inspected at each gauging station every four to eight weeks. The relation between stream surface elevation and stream discharge is evaluated following each discharge measurement at each site and the relationship is adjusted if necessary.

The GAGES database incorporated a QC procedure for delineating the watershed boundaries acquired from the National Hydrography Dataset Plus. The dataset was cross-checked against information from USGS's National Water-Quality Assessment Program. Basin boundaries that were inconsistent across sources were visually compared and manually delineated based on geographical information provided in USGS's Elevation Derivatives for National Applications. Other screening and data quality issues are addressed in the GAGES metadata available at:
http://esapubs.org/archive/ecol/E091/045/metadata.htm.

Analysis

8. Comparability Over Time and Space

All USGS streamflow data have been collected and extensively quality-assured by USGS since the start of data collection. Consistent and well documented procedures have been used for the entire periods of recorded streamflows at all gauges (Corbett et al., 1943; Rantz et al., 1982; Sauer, 2002).

Trends in streamflow over time can be heavily influenced by human activities upstream, such as the construction and operation of dams, discharge of treated wastewater, and land use change. To remove these artificial influences to the extent possible, this indicator relies on a set of reference gauges that were chosen because they represent least-disturbed (though not necessarily completely undisturbed) watersheds. The criteria for selecting reference gauges vary from region to region due to the land use characteristics. This inconsistency means that a modestly impacted gauge in one part of the country (for example, an area with significant agricultural land use) might not have met the data quality standards for another less impacted region. The reference gauge screening process is described in Falcone et al. (2010) and is available in the GAGES metadata at:
http://esapubs.org/archive/ecol/E091/045/metadata.htm.

Analytical methods have been applied consistently over time and space.

9. Sources of Uncertainty

Uncertainty estimates are not available for this indicator as a whole. As for the underlying data, the precision of individual stream gauges varies from site to site. Accuracy depends primarily on the stability of the stage-discharge relationship, the frequency and reliability of stage and discharge measurements, and the presence of special conditions such as ice (Novak, 1985). Accuracy classifications for all USGS gauges for each year of record are available in USGS annual state water data reports. USGS has published a general online reference devoted to the calculation of error in individual stream discharge measurements (Sauer and Meyer, 1992).

10. Sources of Variability

Streamflow can be highly variable over time, depending on the size of the watershed and the factors that influence flow at a gauge. USGS addresses this variability by recording stage several times a day (typically 15-minute to 1-hour intervals) and then computing a daily average streamflow. Streamflow also varies from year to year as a result of variation in precipitation and air temperature. Trend magnitudes computed from Sen slopes provide a robust estimate of linear changes over a period of record, and thus this indicator does not measure decadal cycles or interannual variability in the metric over the time period examined.

While gauges are chosen to represent drainage basins relatively unimpacted by human disturbance, some sites may be more affected by direct human influences (such as land cover and land use change) than others. Other sources of variability include localized factors such as topography, geology, elevation, and natural land cover. Changes in land cover and land use over time can contribute to streamflow trends, though careful selection of reference gauges strives to minimize these impacts.

Although WSCV is largely driven by the timing of the bulk of snow melt in areas with substantial annual snowpack, other factors also will influence WSCV. For instance, a heavy rain event in the winter could result in large volumes of water that shift the timing of the center of volume earlier. Changes over time in the distribution of rainfall during the January–May period could also affect the WSCV date.

11. Statistical/Trend Analysis

The maps in Figures 1, 2, and 3 all show trends over time that have been computed for each gauging station using a Sen slope analysis. Because of uncertainties and complexities in the interpretation of statistical significance, particularly related to the issue of long-term persistence (Cohn and Lins, 2005; Koutsoyiannis and Montanari, 2007), significance of trends is not reported.

12. Data Limitations

Factors that may impact the confidence, application, or conclusions drawn from this indicator are as follows:

1. This analysis is restricted to locations where streamflow is not highly disturbed by human influences, including reservoir regulation, diversions, and land cover change. However, changes in land cover and land use over time could still influence trends in the magnitude and timing of streamflow at some sites.
2. Reference gauges used for this indicator are not evenly distributed throughout the United States, nor are they evenly distributed with respect to topography, geology, elevation, or land cover.

References

Burns, D.A., J. Klaus, and M.R. McHale. 2007. Recent climate trends and implications for water resources in the Catskill Mountain region, New York, USA. J. Hydrol. 336(1–2):155–170.

Cohn, T.A., and H.F. Lins. 2005. Nature's style: Naturally trendy. Geophys. Res. Lett. 32:L23402.

Corbett, D.M., et al. 1943. Stream-gaging procedure: A manual describing methods and practices of the Geological Survey. U.S. Geological Survey Water-Supply Paper 888.

Falcone, J.A., D.M. Carlisle, D.M. Wolock, and M.R. Meador. 2010. GAGES: A stream gage database for evaluating natural and altered flow conditions in the conterminous United States. Ecology 91(2):621.

Helsel, D.R., and R.M. Hirsch. 1992. Statistical methods in water resources. New York, NY: Elsevier.

Hodgkins, G.A., and R.W. Dudley. 2006. Changes in the timing of winter-spring streamflows in eastern North America, 1913–2002. Geophys. Res. Lett. 33:L06402.
http://water.usgs.gov/climate_water/hodgkins_dudley_2006b.pdf.

Koutsoyiannis, D., and A. Montanari. 2007. Statistical analysis of hydroclimatic time series: Uncertainty and insights. Water Resour. Res. 43(5):W05429.

Novak, C.E. 1985. WRD data reports preparation guide. U.S. Geological Survey, Water Resources Division.

Rantz, S.E., et al. 1982. Measurement and computation of streamflow. Volume 1: Measurement of stage and discharge. Volume 2: Computation of discharge. U.S. Geological Survey Water Supply Paper 2175.
http://pubs.usgs.gov/wsp/wsp2175/.

Sauer, V.B. 2002. Standards for the analysis and processing of surface-water data and information using electronic methods. U.S. Geological Survey Water-Resources Investigations Report 01-4044.

Sauer, V.B., and R.W. Meyer. 1992. Determination of error in individual discharge measurements. U.S. Geological Survey Open-File Report 92-144. http://pubs.er.usgs.gov/pubs/ofr/ofr92144.

Sauer, V.B., and D.P. Turnipseed. 2010. Stage measurement at gaging stations. U.S. Geological Survey Techniques and Methods book 3. U.S. Geological Survey. Chap. A7. http://pubs.usgs.gov/tm/tm3-a7.

Turnipseed, D.P., and V.P. Sauer. 2010. Discharge measurements at gaging stations. U.S. Geological Survey Techniques and Methods book 3. U.S. Geological Survey. Chap. A8. http://pubs.usgs.gov/tm/tm3-a8/.

Ragweed Pollen Season

Identification

1. Indicator Description

This indicator describes trends in the annual length of pollen season for ragweed (*Ambrosia* species) at ten North American sites from 1995 to 2011. Ragweed season begins with the shift to shorter daylight after the summer solstice, and it ends in response to cold weather in the fall (i.e., first frost). These constraints suggest that the length of ragweed pollen season is sensitive to climate change by way of changes to fall temperatures.

2. Revision History

December 2011: Indicator developed
May 2012: Indicator updated with data through 2011

Data Sources

3. Data Sources

Data for this indicator come from the National Allergy Bureau. As a part of the American Academy of Allergy, Asthma, and Immunology's (AAAAI's) Aeroallergen Network, the National Allergy Bureau collects pollen data from dozens of stations around the United States. Canadian pollen data originate from Aerobiology Research Laboratories. The data were compiled and analyzed for this indicator by a team of researchers who published a more detailed version of this analysis in 2011, based on data through 2009 (Ziska et al., 2011).

4. Data Availability

EPA acquired data for this indicator from Dr. Lewis Ziska of the U.S. Department of Agriculture, Agricultural Research Service. Dr. Ziska was the lead author of the original analysis published in 2011 (Ziska et al., 2011). He provided an updated version for EPA's indicator, with data through 2011.

Users can access daily ragweed pollen records for each individual U.S. pollen station on the National Allergy Bureau's website at: www.aaaai.org/global/nab-pollen-counts.aspx. *Ambrosia* spp. is classified as a "weed" by the National Allergy Bureau and appears in its records accordingly. Canadian pollen data are not publicly available, but can be purchased from Aerobiology Research Laboratories at: www.aerobiology.ca/products/data.php.

Methodology

5. Data Collection

This indicator is based on daily pollen counts from 10 long-term sampling stations in central North America. Eight sites were in the United States; two sites were in Canada. Sites were selected based on availability of pollen data and nearby weather data (as part of a broader analysis of causal factors) and to represent a variety of latitudes along a roughly north-south transect. Sites were also selected for consistency of altitude and other locational variables that might influence pollen counts.

Data were available from 1995 to 2011 at all sites except for two: Georgetown, Texas (near Austin) had data from 1998 to 2011, and Rogers, Arkansas, had data from 1996 to 2011.

Each station relies on trained individuals to collect air samples. Samples were collected using one of three methods at each counting station:

1. Slide gathering: Blank slides with an adhesive are left exposed to outdoor air to collect airborne samples.
2. Rotation impaction aeroallergen sampler: An automated, motorized device that spins air of a known volume such that airborne particles adhere to a surrounding collection surface.
3. Automated spore sampler from Burkard Scientific: A device that couples a vacuum pump and a sealed rolling tumbler of adhesive paper in a way that records spore samples over time.

Despite differences in sample collection, all sites rely on the human eye to identify and count spores on microscope slides. All of these measurement methods follow standard peer-reviewed protocols. The resulting data sets from AAAAI and Aerobiology Research Laboratories have supported a variety of peer-reviewed studies. Although the sample collection methodologies do not allow for a comparison of total pollen counts across stations that used different methods, the methods are equally sensitive to the appearance of a particular pollen species.

6. Indicator Derivation

By reviewing daily ragweed pollen counts over an entire season, analysts established start and end dates for each location as follows:

- The start date is the point at which 1 percent of the cumulative pollen count for the season has been observed, meaning 99 percent of all ragweed pollen appears after this day.
- The end date is the point at which 99 percent of the cumulative pollen count for the season has been observed.

The duration of pollen season is simply the length of time between the start date and end date.

Two environmental parameters constrain the data used in calculating the length of ragweed season. As a short-day plant, ragweed will not flower before the summer solstice. Furthermore, ragweed is sensitive to frost and will not continue flowering once temperatures dip below freezing (Deen et al., 1998). Because of these two biological constraints, ragweed pollen identified before June 21 or after the first fall frost (based on local weather data) was not included in the analysis.

Once the length of the pollen season was determined for each year and location, a best-fit regression line was calculated for the period from 1995 to 2011. The change in ragweed pollen season (days) from 1995 to 2011 is derived from the slope of this trendline.

Ziska et al. (2011) describe analytical methods in greater detail.

7. Quality Assurance and Quality Control

Pollen counts are determined by trained individuals who follow standard protocols, including procedures for quality assurance and quality control (QA/QC). To be certified as a pollen counter, one must meet various quality standards for sampling and counting proficiency.

Analysis

8. Comparability Over Time and Space

Different stations use different sampling methods, so absolute pollen counts are not comparable across stations. However, because all of the methods are consistent in how they identify the start and end of the pollen season, the season's length data are considered comparable over time and from station to station.

9. Sources of Uncertainty

Error bars for the calculated start and end dates for the pollen season at each site were included in the dataset that was provided to EPA. Identification of the ragweed pollen season start and end dates may be affected by a number of factors, both human and environmental. For stations using optical identification of ragweed samples, the technicians evaluating the slide samples are subject to human error. Further discussion of error and uncertainty can be found in Ziska et al. (2011).

10. Sources of Variability

Wind and rain may impact the apparent ragweed season length. Consistently windy conditions could keep pollen particles airborne for longer periods of time, thereby extending the apparent season length. Strong winds could also carry ragweed pollen long distances from environments with more favorable growing conditions. In contrast, rainy conditions have a tendency to draw pollen out of the air. Extended periods of rain late in the season could prevent what would otherwise be airborne pollen from being identified and recorded.

11. Statistical/Trend Analysis

The indicator relies on a best-fit regression line for each sampling station to determine the change in ragweed pollen season. The 95 percent confidence limits for start and end dates of the pollen season were calculated using Sigmaplot, using the observed data as a basis for the analysis. Changes in season length at six of the 10 stations were deemed to be statistically significant, based on these 95 percent confidence intervals: Saskatoon, Saskatchewan; Winnipeg, Manitoba; Fargo, North Dakota; Minneapolis, Minnesota; LaCrosse, Wisconsin; and Madison, Wisconsin. For further discussion and an earlier version of this significance analysis, see Ziska et al. (2011).

12. Data Limitations

Factors that may impact the confidence, application, or conclusions drawn from this indicator are as follows:

1. This indicator only focuses on 10 stations in the central part of North America. The impacts of climate change on ragweed growth and pollen production could vary in other regions, such as coastal or mountainous areas.
2. This indicator does not describe the extent to which the intensity of ragweed pollen season (i.e., pollen counts) may also be changing.
3. The indicator is sensitive to other factors aside from weather, including the distribution of plant species as well as pests or diseases that impact ragweed or competing species.
4. Although some stations have pollen data dating back to 1973, this indicator only examines trends from 1995 to 2011, based on data availability for the majority of the stations in the analysis.

References

Deen, W., L.A. Hunt, and C.J. Swanton. 1998. Photothermal time describes common ragweed (*Ambrosia artemisiifolia L.*) phenological development and growth. Weed Sci. 46:561–568.

Ziska, L., K. Knowlton, C. Rogers, D. Dalan, N. Tierney, M. Elder, W. Filley, J. Shropshire, L.B. Ford, C. Hedberg, P. Fleetwood, K.T. Hovanky, T. Kavanaugh, G. Fulford, R.F. Vrtis, J.A. Patz, J. Portnoy, F. Coates, L. Bielory, and D. Frenz. 2011. Recent warming by latitude associated with increased length of ragweed pollen season in central North America. PNAS 108:4248–4251.

Length of Growing Season

Identification

1. Indicator Description

This indicator measures the length of the growing season in the contiguous 48 states between 1895 and 2011. The growing season often determines which crops can be grown in an area, as some crops require long growing seasons, while others mature rapidly.

Components of this indicator include:

- Length of growing season in the contiguous 48 states, both nationally (Figure 1) and for the eastern and western halves of the country (Figure 2)
- Timing of the last spring frost and the first fall frost in the contiguous 48 states (Figure 3)

2. Revision History

April 2010: Indicator posted
December 2011: Updated with data through 2010
April 2012: Updated with data through 2011

Data Sources

3. Data Sources

Data were provided by Dr. Kenneth Kunkel of the National Oceanic and Atmospheric Administration's (NOAA's) Cooperative Institute for Climate and Satellites (CICS), who analyzed minimum daily temperature records from weather stations throughout the contiguous 48 states. Temperature measurements come from weather stations in NOAA's Cooperative Observer Program (COOP).

4. Data Availability

EPA obtained the data for this indicator from Dr. Kenneth Kunkel at NOAA CICS. Dr. Kunkel had published an earlier version of this analysis in the peer-reviewed literature (Kunkel et al., 2004), and he provided EPA with an updated file containing growing season data through 2011.

All raw COOP data are maintained by the NOAA's National Climatic Data Center (NCDC). Complete COOP data, embedded definitions, and data descriptions can be downloaded from the Web at: www.ncdc.noaa.gov/doclib/. State-specific data can be found at: www7.ncdc.noaa.gov/IPS/coop/coop.html;jsessionid=312EC0892FFC2FBB78F63D0E3ACF6CBC. There are no confidentiality issues that could limit accessibility, but some portions of the data set might need to be formally requested. Complete metadata for the COOP data set can be found at: www.nws.noaa.gov/om/coop/.

Methodology

5. Data Collection

This indicator focuses on the timing of frosts, specifically the last frost in spring and the first frost in fall. It was developed by analyzing minimum daily temperature records from COOP weather stations throughout the contiguous 48 states.

COOP stations generally measure temperature at least hourly, and they record the minimum temperature for each 24-hour time span. Cooperative observers include state universities, state and federal agencies, and private individuals whose stations are managed and maintained by NOAA's National Weather Service (NWS). Observers are trained to collect data, and the NWS provides and maintains standard equipment to gather these data. The COOP data set represents the core climate network of the United States (Kunkel et al., 2005). Data collected by COOP sites are referred to as U.S. Daily Surface Data or Summary of the Day data.

The study on which this indicator is based includes data from 750 stations in the contiguous 48 states. These stations were selected because they met criteria for data availability; each station had to have less than 10 percent of temperature data missing over the period from 1895 to 2011. For a map of these station locations, see Kunkel et al. (2004). Pre-1948 COOP data were previously only available in hard copy, but were recently digitized by NCDC, thus allowing analysis of more than 100 years of weather and climate data.

Temperature monitoring procedures are described in the full metadata for the COOP data set available at: www.nws.noaa.gov/om/coop/. General information on COOP weather data can be found at: www.nws.noaa.gov/os/coop/what-is-coop.html.

6. Indicator Derivation

For this indicator, the length of the growing season is defined as the period of time between the last frost of spring and the first frost of fall, when the air temperature drops below the freezing point of 32°F. Minimum daily temperature data from the COOP data set were used to determine the dates of last spring frost and first fall frost using an inclusive threshold of 32°F. Methods for producing regional and national trends were designed to weight all regions evenly regardless of station density.

Figure 1 shows trends in the overall length of the growing season, which is the number of days between the last spring frost and the first fall frost. Figure 2 shows trends in the length of growing season for the eastern United States versus the western United States, using 100°W longitude as the dividing line between the two halves of the country. Figure 3 shows trends in the timing of the last spring frost and the first fall frost, also using units of days.

All three figures show the deviation from the 1895–2011 long-term average, which is set at zero for reference. Thus, if spring frost timing in year *n* is shown as -4, it means the last spring frost arrived four days earlier than usual. Note that the choice of baseline period will not affect the shape or the statistical significance of the overall trend; it merely moves the trend up or down on the graph in relation to the point defined as "zero."

To smooth out some of the year-to-year variability and make the results easier to understand visually, all three figures plot 11-year moving averages rather than annual data. EPA chose this averaging period to be consistent with the recommended averaging method used by Kunkel et al. (2004) in an earlier version of this analysis. Each average is plotted at the center of the corresponding 11-year window. For example, the average from 2001 to 2011 is plotted at year 2006. EPA used endpoint padding to extend the 11-year smoothed lines all the way to the ends of the period of record. Per the data provider's recommendation, EPA calculated smoothed values centered at 2007, 2008, 2009, 2010, and 2011 by inserting the 2006–2011 average into the equation in place of the as-yet unreported annual data points for 2012 and beyond. EPA used an equivalent approach at the beginning of the time series.

Kunkel et al. (2004) provide a complete description of the analytical procedures used to determine length of growing season trends. No attempt has been made to represent data outside the contiguous 48 states or to estimate trends before or after the 1895–2011 period.

7. Quality Assurance and Quality Control

NOAA follows extensive quality assurance and quality control (QA/QC) procedures for collecting and compiling COOP weather station data. For documentation of COOP methods, including training manuals and maintenance of equipment, see: www.nws.noaa.gov/os/coop/training.htm. These training materials also discuss QC of the underlying data set. Pre-1948 COOP data were recently digitized from hard copy. Kunkel et al. (2005) discuss QC steps associated with digitization and other factors that might introduce error into the growing season analysis.

The data used in this indicator were carefully analyzed in order to identify and eliminate outlying observations. A value was identified as an outlier if a climatologist judged the value to be physically impossible based on the surrounding values, or if the value of a data point was more than five standard deviations from the station's monthly mean. Readers can find more details on QC analysis for this indicator in Kunkel et al. (2004) and Kunkel et al. (2005).

Analysis

8. Comparability Over Time and Space

Data from individual weather stations were averaged in order to determine national and regional trends in the length of growing season and the timing of spring and fall frosts. To ensure spatial balance, national and regional values were computed using a spatially weighted average, and as a result, stations in low-station-density areas make a larger contribution to the national or regional average than stations in high-density areas.

9. Sources of Uncertainty

Kunkel et al. (2004) present uncertainty measurements for an earlier (but mostly similar) version of this analysis. To test worst-case conditions, Kunkel et al. (2004) computed growing season trends for a thinned-out subset of stations across the country, attempting to simulate the density of the portions of the country with the lowest overall station density. The 95 percent confidence intervals for the resulting trend in length of growing season were ±2 days. Thus, there is very high likelihood that observed changes in growing season are real and not an artifact of sampling error.

10. Sources of Variability

At any given location, the timing of spring and fall frosts naturally varies from year to year as a result of normal variation in weather patterns, multi-year climate cycles such as the El Niño–Southern Oscillation and Pacific Decadal Oscillation, and other factors. This indicator accounts for these factors by applying an 11-year smoothing filter and by presenting a long-term record (more than a century of data). Overall, variability should not impact the conclusions that can be inferred from the trends shown in this indicator.

11. Statistical/Trend Analysis

EPA calculated long-term trends by ordinary least-squares regression to support statements in the "Key Points" text. Kunkel et al. (2004) determined that the overall increase in growing season was statistically significant at a 95 percent confidence level in both the East and the West.

12. Data Limitations

Factors that may impact the confidence, application, or conclusions drawn from this indicator are as follows:

1. Changes in measurement techniques and instruments over time can affect trends. However, these data were carefully reviewed for quality, and values that appeared invalid were not included in the indicator. This indicator only includes data from weather stations that did not have many missing data points.
2. The urban heat island effect can influence growing season data; however, these data were carefully quality-controlled and outlying data points were not included in the calculation of trends.

References

Kunkel, K.E., D.R. Easterling, K. Hubbard, and K. Redmond. 2004. Temporal variations in frost-free season in the United States: 1895–2000. Geophys. Res. Lett. 31:L03201.

Kunkel, K.E., D.R. Easterling, K. Hubbard, K. Redmond, K. Andsager, M.C. Kruk, and M.L. Spinar. 2005. Quality control of pre-1948 Cooperative Observer Network data. J. Atmos. Ocean Tech. 22:1691–1705.

Leaf and Bloom Dates

Identification

1. Indicator Description

This indicator examines the timing of leaf growth and flower blooms for selected plants in the United States between 1900 and 2010. A warming climate can lead to the earlier arrival of spring events, which can cause a variety of impacts on ecosystems and human society.

Components of this indicator include:

- Trends in first leaf dates in the contiguous 48 states since 1900 (Figure 1)
- Trends in first bloom dates in the contiguous 48 states since 1900 (Figure 2)

2. Revision History

April 2010: Indicator posted
December 2011: Updated with data through 2010

Data Sources

3. Data Sources

This indicator is based on leaf and bloom observations that were compiled by the USA National Phenology Network (USA-NPN) and climate data that were provided by the U.S. Historical Climatology Network (USHCN) and other databases maintained by the National Oceanic and Atmospheric Administration's (NOAA's) National Climatic Data Center (NCDC). Data for this indicator were analyzed using an enhanced version of the method described by Schwartz et al. (2006).

4. Data Availability

Phenological Observations

This indicator is based in part on observations of lilac and honeysuckle leaf and bloom dates, to the extent that these observations contributed to the development of models. USA-NPN provides online access to historical phenological observations at: www.usanpn.org/?q=data_main.

Temperature Data

This indicator is based in part on historical daily temperature records, which are publicly available online through NCDC. For example, USHCN data are available online at:
www.ncdc.noaa.gov/oa/climate/research/ushcn/#access, with no confidentiality issues limiting accessibility. Appropriate metadata and "readme" files are appended to the data so that they are discernible for analysis. For example, see:
ftp://ftp.ncdc.noaa.gov/pub/data/ushcn/v2/monthly/readme.txt. Summary data from other sets of weather stations can be obtained from NCDC at: www.ncdc.noaa.gov/oa/ncdc.html.

Model Results

The processed leaf and bloom date data set is not publicly available. EPA obtained the model outputs by contacting Dr. Mark Schwartz at the University of Wisconsin–Milwaukee, who developed the analysis and created the original time series. An earlier version of this analysis appeared in Schwartz et al. (2006); the latest version is currently in press.

Methodology

5. Data Collection

This indicator was developed using models that relate phenological observations (leaf and bloom dates) to weather and climate variables. These models were developed by analyzing the relationships between two types of measurements: 1) observations of the first leaf emergence and the first flower bloom of the season in lilacs and honeysuckles and 2) temperature data. The models were developed using measurements collected throughout the portions of the Northern Hemisphere where lilacs and/or honeysuckles grow, then applied to temperature records from a larger set of stations throughout the contiguous 48 states.

Phenological Observations

First leaf date is defined as the date on which leaves first start to grow beyond their winter bud tips. First bloom date is defined as the date on which flowers start to open. Ground observations of leaf and bloom dates were gathered by government agencies, field stations, educational institutions, and trained citizen scientists; these observations were then compiled by organizations such as the USA-NPN. These types of phenological observations have a long history and have been used to support a wide range of peer-reviewed studies. See Schwartz et al. (2006) and references cited therein for more information about phenological data collection methods.

Temperature Data

Weather data used to construct, validate, and then apply the models—specifically daily maximum and minimum temperatures—were collected from officially recognized weather stations using standard meteorological instruments. These data have been compiled by NCDC databases such as the USHCN and TD3200 Daily Summary of the Day data from other cooperative weather stations. As described in Schwartz et al. (2006), station data were used rather than gridded values, "primarily because of the undesirable homogenizing effect that widely available coarse-resolution grid point data can have on spatial differences, resulting in artificial uniformity of processed outputs..." (Schwartz and Reiter, 2000; Schwartz and Chen, 2002; Menzel et al., 2003). The approximately 600 weather stations were selected according to the following criteria:

- Provide for the best temporal and spatial coverage possible. At some stations, the period of record includes most of the 20th century.
- Have at least 25 of 30 years during the 1981–2010 baseline period, with no 30-day periods missing more than 10 days of data.
- Have sufficient spring–summer warmth to generate valid model output.

For more information on the procedures used to obtain temperature data, see Schwartz et al. (2006) and references cited therein.

6. Indicator Derivation

Daily temperature data and observations of first leaf and bloom dates were used to construct and validate a set of models that relate phenological observations to weather and climate variables (specifically daily maximum and minimum temperatures). These models were developed for the entire Northern Hemisphere and validated at 378 sites in Germany, Estonia, China, and the United States.

Once the models were validated, they were applied to locations throughout the contiguous 48 states using temperature records from 1900 to 2010. Even if actual phenological observations were not collected at a particular station, the models essentially predict phenological behavior based on observed daily maximum and minimum temperatures, allowing the user to estimate the date of first leaf and first bloom for each year at that location. The value of these models is that they can estimate the onset of spring events in locations and time periods where actual lilac and honeysuckle observations are sparse. In the case of this indicator, the models have been applied to a time period that is much longer than most phenological observation records. The models have also been extended to areas of the contiguous 48 states where lilacs and honeysuckles do not actually grow—mainly parts of the South and the West coast where winter is too warm to provide the extended chilling that these plants need in order to bloom the following spring. This step was taken to provide more complete spatial coverage.

This indicator was developed by applying phenological models to several hundred sites in the contiguous 48 states where sufficient weather data have been collected. The exact number of sites varies from year to year depending on data availability (the minimum was 246 sites in 1900; the maximum was 690 sites in 1991).

After running the models, analysts looked at each location and compared the first leaf date and first bloom date in each year with the average leaf date and bloom date for 1981 to 2010, which was established as a "climate normal" or baseline. This step resulted in a data set that lists each station along with the "departure from normal" for each year—measured in days—for each component of the indicator (leaf date and bloom date). Note that 1981 to 2010 represents an arbitrary baseline for comparison, and choosing a different baseline period would shift the observed long-term trends up or down but would not alter the shape, magnitude, or statistical significance of the trends.

EPA obtained a data set listing annual departure from normal for each station, then performed some additional steps to create Figures 1 and 2. For each component of the indicator (leaf date and bloom date), EPA aggregated the data for each year to determine an average departure from normal across all stations. This step involved calculating an unweighted arithmetic mean of all stations with data in a given year. The aggregated annual trend line appears as a thin curve in each figure. To smooth out some of the year-to-year variability, EPA also calculated a nine-year weighted moving average for each component of the indicator. This curve appears as a thick line in each figure, with each value plotted at the center of the corresponding nine-year window. For example, the average from 2000 to 2008 is plotted at year 2004. This nine-year average was constructed using a normal curve weighting procedure that preferentially weights values closer to the center of the window. Weighting coefficients for values 1 through 9, respectively, were as follows: 0.0076, 0.036, 0.1094, 0.214, 0.266, 0.214, 0.1094, 0.036, 0.0076. This procedure was recommended by the authors of Schwartz et al. (2006) as an appropriate way to reduce some of the "noise" inherent in annual phenology data.

EPA used endpoint padding to extend the nine-year smoothed lines all the way to the ends of the period of record. Per the data provider's recommendation, EPA calculated smoothed values centered at 2007, 2008, 2009, and 2010 by inserting the 2006–2010 average into the equation in place of the as-yet unreported annual data points for 2011 and beyond. EPA used an equivalent approach at the beginning of the time series.

For more information on the procedures used to develop, test, and apply the models for this indicator, see Schwartz et al. (2006), McCabe et al. (2011), and references cited therein. This indicator is based on an approach that has been modified slightly since McCabe et al. (2011), in that it uses a fixed start date of January 1 to begin the calculation of warmth accumulation for all station-years, rather than a dynamically calculated start date based on chill hour accumulation, as did the original method. The newest approach is described in Schwartz et al. (in press).

Indicator Development

The 2010 edition of EPA's *Climate Change Indicators in the United States* report presented an earlier version of this indicator based on an analysis published in Schwartz et al. (2006). That analysis was referred to as the Spring Indices (SI). More recently, the team that developed the original Spring Indices has developed an enhanced version of their algorithm, which is referred to as the Extended Spring Indices (SI-x). EPA's indicator has now adopted the SI-x approach. The SI-x represents an extension of the original SI because it can now characterize the timing of spring events in areas where lilacs and honeysuckles do not grow. Additional details about the SI-x are discussed in a paper currently in press.

7. Quality Assurance and Quality Control

Phenological Observations

Quality assurance and quality control (QA/QC) procedures for phenological observations are not readily available.

Temperature Data

Most of the daily maximum and minimum temperature values were evaluated and cleaned to remove questionable values as part of their source development. For example, several papers have been written about the methods of processing and correcting historical climate data for the USHCN. NCDC's website (www.ncdc.noaa.gov/oa/climate/research/ushcn) describes the underlying methodology and cites peer-reviewed publications justifying this approach.

Before applying the model, all temperature data were checked to ensure that no daily minimum temperature value was larger than the corresponding daily maximum temperature value (Schwartz et al., 2006).

Model Results

QA/QC procedures are not readily available regarding the use of the models and processing the results. These models and results have been published in numerous peer-reviewed studies, however, suggesting a high level of QA/QC and review. For more information about the development and application of these models, see Schwartz et al. (2006), McCabe et al. (2011), and the references cited therein.

Analysis

8. Comparability Over Time and Space

Phenological Observations

For consistency, the phenological observations used to develop this indicator were restricted to certain cloned species of lilac and honeysuckle. Using cloned species minimizes the influence of genetic differences in plant response to temperature cues, and it helps to ensure consistency over time and space.

Temperature Data

The USHCN has undergone extensive testing to identify errors and biases in the data and either remove these stations from the time series or apply scientifically appropriate correction factors to improve the utility of the data. In particular, these corrections address changes in the time-of-day of observation, advances in instrumentation, and station location changes.

Homogeneity testing and data correction methods are described in more than a dozen peer-reviewed scientific papers by NCDC. Data corrections were developed to specifically address potential problems in trend estimation of the rates of warming or cooling in the USHCN. Balling and Idso (2002) compare the USHCN data with several surface and upper-air data sets and show that the effects of the various USHCN adjustments produce a significantly more positive, and likely spurious, trend in the USHCN data. In contrast, a subsequent analysis by Vose et al. (2003) found that USHCN station history information is reasonably complete and that the bias adjustment models have low residual errors.

Further analysis by Menne et al. (2009) suggests that:

> ...the collective impact of changes in observation practice at USHCN stations is systematic and of the same order of magnitude as the background climate signal. For this reason, bias adjustments are essential to reducing the uncertainty in U.S. climate trends. The largest biases in the HCN are shown to be associated with changes to the time of observation and with the widespread changeover from liquid-in-glass thermometers to the maximum minimum temperature sensor (MMTS). With respect to [USHCN] Version 1, Version 2 trends in maximum temperatures are similar while

minimum temperature trends are somewhat smaller because of an apparent overcorrection in Version 1 for the MMTS instrument change, and because of the systematic impact of undocumented station changes, which were not addressed [in] Version 1.

USHCN Version 2 represents an improvement in this regard.

Some observers have expressed concerns about other aspects of station location and technology. For example, Watts (2009) expresses concern that many U.S. weather stations are sited near artificial heat sources such as buildings and paved areas, potentially biasing temperature trends over time. In response to these concerns, NOAA analyzed trends for a subset of stations that Watts had determined to be "good or best," and found the temperature trend over time to be very similar to the trend across the full set of USHCN stations (www.ncdc.noaa.gov/oa/about/response-v2.pdf). While it is true that many stations are not optimally located, NOAA's findings support the results of an earlier analysis by Peterson (2006) that found no significant bias in long-term trends associated with station siting once NOAA's homogeneity adjustments have been applied.

Model Results

The same model was applied consistently over time and space. This indicator generalizes results over space by averaging station-level departures from normal in order to determine the aggregate departure from normal for each year. This step uses a simple unweighted arithmetic average, which is appropriate given the national scale of this indicator and the large number of weather stations spread across the contiguous 48 states.

9. Sources of Uncertainty

Error estimates are not readily available for the underlying temperature data upon which this indicator is based. It is generally understood that uncertainties in the temperature data increase as one goes back in time, as there are fewer stations early in the record. However, these uncertainties are not sufficient to mislead the user about fundamental trends in the data.

In aggregating station-level "departure from normal" data into an average departure for each year, EPA calculated the standard error of each component of the indicator (leaf date and bloom date) in each year. For both components, standard errors range from 0.2 days to 0.8 days, depending on the year.

Schwartz et al. (2006) provide error estimates for the models as well as for similar indicators considered across the entire Northern Hemisphere. The use of modeled data should not detract from the conclusions that can be inferred from the indicator. These models have been extensively tested and refined over time and space such that they offer good certainty.

10. Sources of Variability

Temperatures naturally vary from year to year, which can strongly influence leaf and bloom dates. To smooth out some of the year-to-year variability, EPA calculated a nine-year weighted moving average for each component of this indicator.

11. Statistical/Trend Analysis

Statistical testing of individual station trends suggests that many of these trends are not significant within the contiguous 48 states. Other studies (e.g., Schwartz et al., 2006) have come to similar conclusions, finding that trends in the earlier onset of spring at individual stations are much stronger in Canada and parts of Eurasia than they are in the contiguous 48 states. In part as a result of these findings, this EPA indicator focuses on aggregate trends across the contiguous 48 states, which should be more statistically robust than individual station trends. However, the aggregate trends still are not statistically significant ($p < 0.05$) over the entire period of record, based on a simple t-test.

12. Data Limitations

Factors that may impact the confidence, application, or conclusions drawn from this indicator are as follows:

1. Plant phenological events are studied using several data collection methods, including satellite images, models, and direct observations. The use of varying data collection methods in addition to the use of different phenological indicators (such as leaf or bloom dates for different types of plants) can lead to a range of estimates of the arrival of spring.
2. Climate is not the only factor that can affect phenology. Observed variations can also reflect plant genetics, changes in the surrounding ecosystem, and other factors. This indicator minimizes genetic influences by relying on cloned plant species, however (that is, plants with no genetic differences).

References

Balling, Jr., R.C., and C.D. Idso. 2002. Analysis of adjustments to the United States Historical Climatology Network (USHCN) temperature database. Geophys. Res. Lett. 29(10):1387.

McCabe, G.J., T.R. Ault, B.I. Cook, J.L. Betancourt, and M.D. Schwartz. 2011. Influences of the El Niño Southern Oscillation and the Pacific Decadal Oscillation on the timing of the North American spring. Int. J. Climatol. (online).

Menne, M.J., C.N. Williams, Jr., and R.S. Vose. 2009. The U.S. Historical Climatology Network monthly temperature data, version 2. Bull. Am. Meteorol. Soc. 90:993-1107. ftp://ftp.ncdc.noaa.gov/pub/data/ushcn/v2/monthly/menne-etal2009.pdf.

Menzel, A., F. Jakobi, R. Ahas, et al. 2003. Variations of the climatological growing season (1051–2000) in Germany compared to other countries. Int. J. Climatol. 23:793–812.

Peterson, T.C. 2006. Examination of potential biases in air temperature caused by poor station locations. Bull. Am. Meteorol. Soc. 87:1073–1080. http://journals.ametsoc.org/doi/pdf/10.1175/BAMS-87-8-1073.

Schwartz, M.D., and X. Chen. 2002. Examining the onset of spring in China. Clim. Res. 21:157–164.

Schwartz, M.D., and B.E. Reiter. 2000. Changes in North American spring. Int. J. Climatol. 20:929–932.

Schwartz, M.D., R. Ahas, and A. Aasa. 2006. Onset of spring starting earlier across the Northern Hemisphere. Glob. Chang. Biol. 12:343–351.

Schwartz, M.D., T.R. Ault, and J.L. Betancourt. In press. Spring onset variations and trends in the continental United States: Past and regional assessment using temperature-based indices. Int. J. Climatol.

Vose, R.S., C.N. Williams, Jr., T.C. Peterson, T.R. Karl, and D.R. Easterling. 2003. An evaluation of the time of observation bias adjustment in the U.S. Historical Climatology Network. Geophys. Res. Lett. 30(20):2046.

Watts, A. 2009. Is the U.S. surface temperature record reliable? The Heartland Institute. http://wattsupwiththat.files.wordpress.com/2009/05/surfacestationsreport_spring09.pdf.

Bird Wintering Ranges

Identification

1. Indicator Description

This indicator examines changes in the winter ranges of North American birds from the winter of 1966–1967 to 2005. Changes in climate can affect ecosystems by influencing animal behavior and distribution. Birds are a particularly strong indicator of environmental change for several reasons described in the indicator text.

Components of this indicator include:

- Shifts in the latitude of winter ranges of North American birds over the past half-century (Figure 1)
- Shifts in the distance to the coast of winter ranges of North American birds over the past half-century (Figure 2)

2. Revision History

April 2010: Indicator posted

Data Sources

3. Data Sources

This indicator is based on data collected by the annual Christmas Bird Count (CBC), managed by the National Audubon Society. Data used in this indicator are collected by citizen scientists who systematically survey certain areas and identify and count common bird species. The CBC has been in operation since 1900, but data used in this indicator begin in winter 1966–1967.

4. Data Availability

Complete CBC data are available in both print and electronic formats. Historical CBC data are available in print in the following periodicals: *Audubon Field Notes*, *American Birds*, and *Field Notes*. Annual publications of CBC data were made available beginning in 1998. Additionally, historical, current year, and annual summary CBC data are available online at: http://birds.audubon.org/christmas-bird-count. Descriptions of data are available with the data queried online. The appendix to National Audubon Society (2009) provides 40-year trends for each species, but not the full set of data by year. EPA obtained the complete data set for this indicator directly from the National Audubon Society.

A similar analysis is available from an interagency consortium at: www.stateofthebirds.org/2010.

Methodology

5. Data Collection

This indicator is based on data collected by the annual CBC, managed by the National Audubon Society. Data used in this indicator are collected by citizen scientists who systematically survey certain areas and identify and count common bird species. Although the indicator relies on human observation rather than precise measuring instruments, the people who collect the data are skilled observers who follow strict protocols that are consistent across time and space. These data have supported many peer-reviewed studies, a list of which can be found on the National Audubon Society's website at: http://web4.audubon.org/bird/cbc/biblio.html.

Bird surveys take place each year in approximately 2,000 different locations throughout the contiguous 48 states and the southern portions of Alaska and Canada. All local counts take place between December 14 and January 5 of each winter. Each local count takes place over a 24-hour period in a defined "count circle" that is 15 miles in diameter. A variable number of volunteer observers separate into field parties, which survey different areas of the count circle and tally the total number of individuals of each species observed (National Audubon Society, 2009). This indicator covers 305 bird species, which are listed in Appendix 1 of National Audubon Society (2009). These species were included because they are widespread and they met specific criteria for data availability.

The entire study description, including sampling methods and analyses performed, can be found in National Audubon Society (2009) and references therein. Information on this study is also available on the National Audubon Society website at: http://birdsandclimate.audubon.org/index.html. For additional information on CBC survey design and methodologies, see the technical reports listed at: www.audubon.org/bird/cbc/biblio.html.

6. Indicator Derivation

At the end of the 24-hour observation period, each count circle tallies the total number of individuals of each species seen in the count circle. Audubon scientists then run the data through several levels of analysis and quality control to determine final count numbers from each circle and each region. Population trends over the 40-year period of this indicator and annual indices of abundance were estimated for the entire survey area with hierarchical models in a Bayesian analysis using Markov chain Monte Carlo techniques (National Audubon Society, 2009). Data processing steps also include corrections for different levels of effort—for example, if some count circles had more observers and more person-hours of effort than others.

This indicator is based on the center of abundance for each species, which is the center of the population distribution at any point in time. In terms of latitude, half of the individuals in the population live north of the center of abundance and the other half live to the south. Similarly, in terms of longitude, half of the individuals live west of the center of abundance, and the other half live to the east. The center of abundance is a common way to characterize the general location of a population. For example, if a population were to shift generally northward, the center of abundance would be expected to shift northward as well.

This indicator examines the center of abundance from two perspectives:

- Latitude—testing the hypothesis that bird populations are moving northward along with the observed rise in overall temperatures throughout North America.
- Distance from coast—testing the hypothesis that bird populations are able to move further from the coast as a generally warming climate moderates the inland temperature extremes that would normally occur in the winter.

This indicator reports the position of the center of abundance for each year, relative to the position of the center of abundance in 1966 (winter 1966–1967), averaged across all 305 species. No attempt was made to generate estimates outside the surveyed area. The indicator does not include northern Alaska or Canada because data for these areas were too sparse to support meaningful trend analysis. No attempt was made to estimate trends prior to 1966 (i.e., prior to the availability of complete spatial coverage and standardized methods), and no attempt was made to project trends into the future.

The entire study description, including analyses performed, can be found in National Audubon Society (2009) and references therein. Information on this study is also available on the National Audubon Society website at: http://birdsandclimate.audubon.org/index.html.

7. Quality Assurance and Quality Control

As part of the overall data compilation effort, Audubon scientists have performed several statistical analyses to ensure that potential error and variability are adequately addressed. QA/QC procedures are described in National Audubon Society (2009) and in a variety of methodology reports listed at: www.audubon.org/bird/cbc/biblio.html.

Analysis

8. Comparability Over Time and Space

The CBC has been in operation since 1900, but data used in this indicator begin in winter 1966–1967. The National Audubon Society chose this start date to ensure sufficient sample size throughout the survey area as well as consistent methods, as the CBC design and methodology have remained generally consistent since the 1960s. All local counts take place between December 14 and January 5 of each winter, and they follow consistent methods regardless of the location.

9. Sources of Uncertainty

The sources of uncertainty in this indicator have been analyzed, quantified, and accounted for to the extent possible. The statistical significance of the trends suggests that the conclusions one might draw from this indicator are robust.

One potential source of uncertainty in these data is uneven effort among count circles. Various studies that discuss the best ways to account for this source of error have been published in peer-reviewed journals. Link and Sauer (1999) describe the methods that Audubon used to account for variability in effort.

10. Sources of Variability

Rare or difficult-to-observe bird species could lead to increased variability. For this analysis, the National Audubon Society included only 305 widespread birds that met criteria for abundance and the availability of data to enable the detection of meaningful trends.

11. Statistical/Trend Analysis

Appendix 1 of National Audubon Society (2009) documents the statistical significance of trends in the wintering range for each species included in this indicator. National Audubon Society (2009) also presents the statistical significance of each of the aggregate trends (northward distance and distance from the coast across all 305 species) and discusses the uncertainty of these trends. Based on ordinary least-squares regression, the average latitudinal center of abundance shifted significantly to the north by 34.8 miles ($p<0.0001$) over the period of interest. Populations shifted inward from the coast by an average of 20.5 miles ($p<0.0001$).

12. Data Limitations

Factors that may impact the confidence, application, or conclusions drawn from this indicator are as follows:

1. Many factors can influence bird ranges, including food availability, habitat alteration, and interactions with other species. Some of the birds covered in this indicator might have moved northward or inland for reasons other than changing temperatures.
2. This indicator does not show how responses to climate change vary among different types of birds. For example, National Audubon Society (2009) found large differences between coastal birds, grassland birds, and birds adapted to feeders, which all have varying abilities to adapt to temperature changes. This Audubon report also shows the large differences between individual species—some of which moved hundreds of miles while others did not move significantly at all.
3. Some data variations are caused by differences between count circles, such as inconsistent level of effort by volunteer observers, but these differences are carefully corrected in Audubon's statistical analysis.
4. While observers attempt to identify and count every bird observed during the 24-hour observation period, rare and nocturnal species may be undersampled.

References

Link, W.A., and J.R. Sauer. 1999. Controlling for varying effort in count surveys: An analysis of Christmas Bird Count data. J. Agric. Biol. Envir. S. 4:116–125.

National Audubon Society. 2009. Northward shifts in the abundance of North American birds in early winter: a response to warmer winter temperatures? www.audubon.org/bird/bacc/techreport.html.

Heat-Related Deaths

Identification

1. Indicator Description

Extreme heat events have become more frequent in most of North America in recent decades (see the High and Low Temperatures indicator), and these events can be associated with increases in heat-related deaths.

Components of this indicator include:

- The rate of U.S. deaths between 1979 and 2009 for which heat was classified on death certificates as the underlying (direct) cause (Figure 1, orange line)
- The rate of U.S. deaths between 1999 and 2009 for which heat was classified as either the underlying cause or a contributing factor (Figure 1, blue line)

2. Revision History

April 2010: Indicator posted
December 2011: Updated with data through 2007; added contributing factors analysis to complement the existing time series
August 2012: Updated with data through 2009; converted the measure from counts to crude rates; added example figure.

Data Sources

3. Data Sources

This indicator is based on data from the U.S. Centers for Disease Control and Prevention's (CDC's) National Vital Statistics System (NVSS), which compiles information from death certificates for nearly every death in the United States. The NVSS is the most comprehensive source of mortality data for the population of the United States. The U.S. Centers for Disease Control and Prevention (CDC) provided analysis of NVSS data.

Mortality data for the illustrative example figure came from CDC's National Center for Health Statistics (NCHS). The estimate of deaths in excess of the average daily death rate is from the National Research Council's report on climate stabilization targets (NRC, 2011), which cites the peer-reviewed publication Kaiser et al. (2007).

For reference, the illustrative example also shows daily maximum temperature data from the weather station at the Chicago O'Hare International Airport (GHCND:USW00094846).

4. Data Availability

Underlying Causes

The long-term trend line (1979–2009) is based on CDC's Compressed Mortality File, which can be accessed through the CDC WONDER online database at: http://wonder.cdc.gov/mortSQL.html (CDC, 2012a). CDC WONDER provides free public access to mortality statistics, allowing users to query data for the nation as a whole or data broken down by state or region, demographic group (age, sex, race), or International Classification of Diseases (ICD) code. Users can obtain the data for this indicator by accessing CDC WONDER and querying the ICD codes listed in Section 5 for the entire U.S. population.

Underlying and Contributing Causes

The 1999–2009 trend line is based on an analysis developed by the National Environmental Public Health Tracking (EPHT) Program, which CDC coordinates. Monthly totals by state are available online at: http://ephtracking.cdc.gov/showIndicatorPages.action. CDC staff from the National Center for Environmental Health (NCEH) EPHT branch provided national totals to EPA (CDC, 2012b). Users can query underlying and contributing causes of death through CDC WONDER's Multiple Causes of Death file (http://wonder.cdc.gov/mcd.html), but EPHT performed additional steps that cannot be recreated through the publicly available data portal (see Section 6).

Death Certificates

Individual-level data (i.e., individual death certificates) are not publicly available due to confidentiality issues.

Chicago Heat Wave Example

Data for the example figure are based on CDC's Compressed Mortality File, which can be accessed through the CDC WONDER online database at: www.cdc.gov/nchs/data_access/cmf.htm. The analysis was obtained from Kaiser et al. (2007). Daily maximum temperature data for 1995 from the Chicago O'Hare International Airport weather station are available from the National Oceanic and Atmospheric Administration's (NOAA's) National Climatic Data Center (NCDC) at: www.ncdc.noaa.gov/oa/climate/stationlocator.html.

Methodology

5. Data Collection

This indicator is based on causes of death as reported on death certificates. A death certificate typically provides space to designate an immediate cause of death along with up to 20 contributing causes, one of which will be identified as the underlying cause of death. The World Health Organization (WHO) defines the underlying cause of death as "the disease or injury which initiated the train of events leading directly to death, or the circumstances of the accident or violence which produced the fatal injury."

Causes of death are certified by a physician, medical examiner, or coroner, and are classified according to a standard set of codes called the ICD. Deaths for 1979 through 1998 are classified using the Ninth Revision of ICD (ICD-9). Deaths for 1999 and beyond are classified using the Tenth Revision (ICD-10).

Although causes of death rely to some degree on the judgment of the physician, medical examiner, or coroner, the "measurements" for this indicator are expected to be generally reliable based on the medical knowledge required of the "measurer" and the use of a standard classification scheme based on widely accepted scientific definitions. When more than one cause or condition is entered, the underlying cause is determined by the sequence of conditions on the certificate, provisions of the ICD, and associated selection rules and modifications.

Mortality data are collected for the entire population and, therefore, are not subject to sampling design error. For virtually every death that occurs in the United States, a physician, medical examiner, or coroner certifies the causes of death on an official death certificate. State registries collect these death certificates and report causes of death to the NVSS. NVSS's shared relationships, standards, and procedures form the mechanism by which the CDC collects and disseminates the nation's official vital statistics.

Standard forms for the collection of data and model procedures for the uniform registration of death events have been developed and recommended for state use through cooperative activities of the states and CDC's NCHS. All states collect a minimum data set specified by NCHS, including underlying causes of death. CDC has published procedures for collecting vital statistics data (CDC, 1995).

This indicator excludes deaths to foreign residents and deaths to U.S. residents who died abroad.

General information regarding data collection procedures can be found in the Model State Vital Statistics Act and Regulations (CDC, 1995). For additional documentation on the CDC WONDER database (EPA's data source for part of this indicator) and its underlying sources, see: http://wonder.cdc.gov/wonder/help/cmf.html.

CDC has posted a recommended standard certificate of death online at: www.cdc.gov/nchs/data/dvs/DEATH11-03final-ACC.pdf. For a complete list and description of the ICD codes used to classify causes of death, see: www.who.int/classifications/icd/en.

Chicago Heat Wave Example

The mortality dataset shown in the example figure includes the entire Standard Metropolitan Statistical Area for Chicago, a region that contains Cook County plus a number of counties in Illinois and Indiana, from June 1 to August 31, 1995.

In the text box above the example figure, values reflect data from Cook County only. The number of deaths classified as "heat-related" on Cook County death certificates between July 11 and July 27, 1995, was reported to CDC by the Cook County Medical Examiner's Office. More information is available in CDC's Morbidity and Mortality Weekly Report (www.cdc.gov/MMWR/preview/mmwrhtml/00038443.htm). Deaths in excess of the average daily death rate for Cook County were determined from death certificates obtained from the Illinois Department of Public Health (Kaiser et al., 2007).

6. Indicator Derivation

This indicator reports the annual rate of deaths per million population that have been classified with ICD codes related to exposure to natural sources of heat. The NVSS collects data on virtually all deaths that occur in the United States, meaning the data collection mechanism already covers the entire target population. Thus, it was not necessary to extrapolate the results on a spatial or population basis. No attempt has been made to reconstruct trends prior to the onset of comprehensive data collection, and no attempt has been made to project data forward into the future.

Underlying Causes

The long-term trend line in Figure 1 reports the rate of deaths per year for which the underlying cause had one of the following ICD codes:

- ICD-9 code E900: "excessive heat—hyperthermia"—specifically subpart E900.0: "due to weather conditions."
- ICD-10 code X30: "exposure to excessive natural heat—hyperthermia."

This component of the indicator is reported for the entire year. EPA developed this analysis based on the publicly available data compiled by CDC WONDER. EPA chose to use crude death rates rather than death counts because rates account for changes in total population over time. Population figures are obtained from CDC WONDER.

Underlying and Contributing Causes

The "underlying and contributing causes" trend line in Figure 1 reports the rate of deaths for which either the underlying cause or the contributing causes had one or more of the following ICD codes:

- ICD-10 code X30: "exposure to excessive natural heat—hyperthermia."
- ICD-10 codes T67.0 through T67.9: "effects of heat and light." Note that the T67 series is only used for contributing causes—never for the underlying cause.

To reduce the chances of including deaths that were incorrectly classified, EPHT did not count the following deaths:

- Deaths occurring during colder months (October through April). Thus, the analysis is limited to May–September.
- Any deaths for which the ICD-10 code W92, "exposure to excessive heat of man-made origin," appears in any cause field. This step removes certain occupational-related deaths.

Foreign residents were excluded. EPHT obtained death counts directly from NVSS, rather than using the processed data available through CDC WONDER. EPHT has not yet applied its methods to data prior to 1999. For a more detailed description of EPHT's analytical methods, see the indicator documentation at: http://ephtracking.cdc.gov/showIndicatorPages.action. Crude death rates were calculated in the same manner as with the underlying causes time series.

Chicago Heat Wave Example

The authors of Kaiser et al. (2007) determined that the Chicago area had 692 deaths in excess of the background death rate between June 21 and August 10, 1995. This analysis excluded deaths from accidental causes but included 183 deaths from "mortality displacement," which refers to a decrease in the deaths of individuals who would have died during this period in any case but whose deaths were accelerated by a few days due to the heat wave. This implies that the actual number of excess deaths during the period of the heat wave itself (July 11–27) was higher than 692, but was compensated for by reduced daily death rates in the week after July 27. Thus the value for excess deaths in Cook County for the period of July 11–27 is reported as approximately 700 in the text box above the example figure.

7. Quality Assurance and Quality Control

Vital statistics regulations have been developed to serve as a detailed guide to state and local registration officials who administer the NVSS. These regulations provide specific instructions to protect the integrity and quality of the data collected. This quality assurance information can be found in CDC (1995).

For the "underlying and contributing causes" component of this indicator, extra steps have been taken to remove certain deaths that could potentially reflect a misclassification (see Section 6). These criteria generally excluded only a small number of deaths.

Analysis

8. Comparability Over Time and Space

When plotting the data, EPA inserted a break in the line between 1998 and 1999 to reflect the transition from ICD-9 codes to ICD-10 codes. The change in codes makes it difficult to accurately compare pre-1999 data with data from 1999 and later. Otherwise, all methods have been applied consistently over time and space. ICD codes allow physicians and other medical professionals across the country to use a standard scheme for classifying causes of deaths.

9. Sources of Uncertainty

Uncertainty estimates are not available for this indicator. Because statistics have been gathered from virtually the entire target population (i.e., all deaths in a given year), these data are not subject to the same kinds of errors and uncertainties that would be inherent in a probabilistic survey or other type of representative sampling program.

Some uncertainty could be introduced as a result of the professional judgment required of the medical professionals filling out the death certificates, which could potentially result in misclassification or underreporting in some number of cases—probably a small number of cases, but still worth noting.

10. Sources of Variability

There is substantial year-to-year variability within the data, due in part to the influence of a few large events. Many of the spikes apparent in Figure 1 can be attributed to specific severe heat waves occurring in large urban areas.

11. Statistical/Trend Analysis

This indicator does not report on the slope of the apparent trends in heat-related deaths, nor does it calculate the statistical significance of these trends.

12. Data Limitations

Factors that may impact the confidence, application, or conclusions drawn from this indicator are as follows:

1. It has been well-documented that many deaths associated with extreme heat are not identified as such by the medical examiner and might not be correctly coded on the death certificate. In many cases, they might just classify the cause of death as a cardiovascular or respiratory disease. They might not know for certain whether heat was a contributing factor, particularly if the death did not occur during a well-publicized heat wave. By studying how daily death rates vary with temperature in selected cities, scientists have found that that extreme heat contributes to far more deaths than the official death certificates would suggest (Medina-Ramón and Schwartz, 2007). That is because the stress of a hot day can increase the chance of dying from a heart attack, other heart conditions, and respiratory diseases such as pneumonia (Kaiser et al., 2007). These causes of death are much more common than heat-related illnesses such as heat stroke. Thus, this indicator very likely underestimates the number of deaths caused by exposure to heat. However, it does serve as a reportable national measure of deaths attributable to heat.
2. ICD-9 codes were used to specify underlying cause of death for the years 1979 to 1998. Beginning in 1999, cause of death was specified with ICD-10 codes. The two revisions differ substantially, so data from before 1999 cannot easily be compared with data from 1999 and later.
3. The fact that a death is classified as "heat-related" does not mean that high temperatures were the only factor that caused the death. Pre-existing medical conditions can greatly increase an individual's vulnerability to heat.
4. Heat waves are not the only factor that can affect trends in "heat-related" deaths. Other factors include the vulnerability of the population, the extent to which people have adapted to higher temperatures, the local climate and topography, and the steps people have taken to manage heat emergencies effectively.
5. Heat response measures can make a big difference in death rates. Response measures can include early warning and surveillance systems, air conditioning, health care, public education, infrastructure standards, and air quality management. For example, after a 1995 heat wave, the City of Milwaukee developed a plan for responding to extreme heat conditions in the future. During the 1999 heat wave, this plan cut heat-related deaths nearly in half compared with what was expected (Weisskopf et al., 2002).

References

CDC (U.S. Centers for Disease Control and Prevention). 1995. Model State Vital Statistics Act and Regulations (revised April 1995). DHHS publication no. (PHS) 95-1115. www.cdc.gov/nchs/data/misc/mvsact92aacc.pdf.

CDC (U.S. Centers for Disease Control and Prevention). 2012a. CDC Wide-ranging Online Data for Epidemiologic Research (WONDER). Compressed mortality file, underlying cause of death. 1999–2009 (with ICD-10 codes) and 1979–1998 (with ICD-9 codes). Accessed August 2012. http://wonder.cdc.gov/mortSQL.html.

CDC (U.S. Centers for Disease Control and Prevention). 2012b. Indicator: Heat-related mortality. National Center for Health Statistics. Annual national totals provided by National Center for Environmental Health staff in August 2012. http://ephtracking.cdc.gov/showIndicatorPages.action.

Kaiser, R., A. Le Tertre, J. Schwartz, C.A. Gotway, W.R. Daley, and C.H. Rubin. 2007. The effect of the 1995 heat wave in Chicago on all-cause and cause-specific mortality. Am. J. Public Health 97(Supplement 1):S158–S162.

Medina-Ramón, M., and J. Schwartz. 2007. Temperature, temperature extremes, and mortality: A study of acclimatization and effect modification in 50 U.S. cities. Occup. Envi. Med. 64(12):827–833.

NRC (National Research Council). 2011. Climate stabilization targets: Emissions, concentrations, and impacts over decades to millennia. Washington, DC: National Academies Press.

Weisskopf, M.G., H.A. Anderson, S. Foldy, L.P. Hanrahan, K. Blair, T.J. Torok, and P.D. Rumm. 2002. Heat wave morbidity and mortality, Milwaukee, Wis, 1999 vs. 1995: An improved response? Am. J. Public Health 92:830–833.

www.ingramcontent.com/pod-product-compliance
Lightning Source LLC
Chambersburg PA
CBHW080638180526
45168CB00008B/3216